# Molecular Biology
# Biochemistry and Biophysics

# Molekularbiologie
# Biochemie und Biophysik

# 6

Springer-Verlag New York Inc. 1969

F. Egami · K. Nakamura

# Microbial Ribonucleases

With 5 Figures

Springer-Verlag New York Inc. 1969

*Dr. Fujio Egami*
Professor of Biochemistry

*Dr. Keiko Nakamura*

Department of Biophysics and Biochemistry, Faculty of Science
The University of Tokyo, Hongo, Tokyo, Japan

ISBN 978-3-642-87500-7          ISBN 978-3-642-87498-7 (eBook)
DOI 10.1007/978-3-642-87498-7

Title No. 3806

# Table of Contents

# Abbreviations

| | |
|---|---|
| RNA | Ribonucleic acid |
| DNA | Deoxyribonucleic acid |
| RNase | Ribonuclease |
| PDase | Phosphodiesterase |
| PMase | Phosphomonoesterase |
| sRNA | Soluble or transfer RNA |
| mRNA | Messenger RNA |
| rRNA | Ribosomal RNA |
| poly X | Linear 3'-5'-polymer of nucleotide X |
| Xp | Nucleoside 3'-phosphate |
| pX | Nucleoside 5'-phosphate |
| X>p | Nucleoside 2',3'-cyclic phosphate |

Chapter I

# Introduction

Studies on microbial RNases began in 1924 when Noguchi found nucleic acid degrading enzymes in Takadiastase, a commercially prepared digest from *Aspergillus oryzae*. In 1935 Otani reported the presence of enzymes degrading yeast RNA in fungi (*Aspergillus sp.* etc.).

In 1948, McCarty studied the nuclease activity in 36 strains of group A hemolytic Streptococci and indicated the release of both RNase and DNase from Streptococcal cells during growth in culture media. As is well known, these nucleases (Tillett et al., 1948) are applied in the treatment of inflammatory exudates.

In 1952, Muggleton and Webb found RNase in the culture medium of Actinomyces (strain A), and they suggested that the ability of the medium to render killed Gram-positive cells Gram-negative may be due to the RNase. In the same year Pardee and his colleagues (1952 a, b) found RNase activity in *E. coli* (strain B) infected by bacteriophage $T_2r$, $T_3$ etc.

Kuninaka (1954) found ribonuclease in culture filtrates of *Aspergillus oryzae*. Saruno (1956) purified the enzyme from culture filtrates of *Aspergillus oryzae* by employing ammonium sulphate fractionation and Rivanol treatment, and reported the existence of an enzyme which differed from pancreatic ribonuclease because it hydrolyzed the purine moiety of RNA to 3'-mononucleotides. Masui et al. (1956) were also able to show enzyme activities of both ribonuclease and desoxyribonuclease in culture filtrates of *Escherichia coli*, *Bacillus subtilis*, and seven strains of pathogenic bacteria. Cunningham et al. (1956) found nucleases in culture filtrates of *Micrococcus pyogenes*.

The studies so far cited were only on the occurrence of nucleic acid-degrading enzymes in microorganisms and did not cover the specificities of these enzymes.

On the other hand, Sato and Egami (1957) found two RNases designated as RNases $T_1$ and $T_2$ in Takadiastase and showed that the former preferentially attacked the bonds next to guanylic acid and formed 3'-guanylic acid via 2',3'-cyclic guanylic acid while the latter preferentially attacked the adenylic acid phosphodiester bonds in RNA.

Thus for the first time RNases with different specificities from pancreatic RNase were demonstrated in a microorganism, and it was suggested that these enzymes might be useful for studies on RNA such as for the determination of its nucleotide sequences (Sato-Asano and Egami, 1960).

The above finding, together with the use of certain 5'-ribonucleotides as flavouring substances, stimulated a systematic search for specific RNases and nucleases producing 5'-nucleotides. Thus, as shown in the "References to Chapter III", much literature has been published on this subject since 1960.

The reasons why RNase is considered to be an attractive subject for biochemical and biological studies may be as follows:

*1. RNases as enzyme proteins.* In the field of protein chemistry, RNases have been used as the most suitable enzymes for the elucidation of the relationship between the structure and function of enzymes, because it is rather easy to purify them and analyze their primary and secondary structures since they are rather small and stable. Moreover, RNases are very suitable proteins for comparative biochemical studies as they are widely distributed in animals, plants, and microorganisms.

*2. Physiological role of RNases.* RNase must play an important role in the metabolism of RNA and may even take part in the mechanism controlling nucleic acid metabolism. Taking into account the essential function of RNA in living matter, a knowledge of the reactions catalyzed by RNase may give some clues as to the mechanisms and controls of all the reactions in organisms.

*3. Applications.*

a) *Analysis of the nucleotide sequence in RNA.* When the specificity and mode of action of an RNase are well established, it is an excellent tool for studying the structure, nucleotide sequence and nearest neighbour frequencies of RNA in combination with exonucleases and chemical methods. This analysis is facilitated by use of an RNase which is specific for only one kind of nucleotide other than guanyloribonucleotide, and so such enzymes must be found by a survey of the enzymes in a wide range of organisms.

b) *Preparation of uucleotides and oligonucleotides..* After enzymatic hydrolysis of RNA, nucleotides and oligonucleotides of various sizes and nucleotide sequences can be obtained by fractionating the hydrolyzate by column chromatography. Moreover, some RNases are known to catalyze the synthesis of oligonucleotides with definite nucleotide sequences from nucleoside 2′,3′-cyclic phosphate (SATO-ASANO, 1960; HAYASHI and EGAMI, 1963).

c) *Application to biological studies.* RNase may be applied in various cytological studies and biological analyses. It can be used for the elimination of contaminating RNA from DNA and proteins. It may be used as a reagent for the removal of RNA from a cell-free system in order to reveal the roles of RNA in protein synthesis.

As mentioned above, RNases should be studied from various aspects. For this purpose, microbial RNases seem to be the most suitable enzymes because of the simplicity of the whole cell system.

The authors would like to discuss here both the chemical and biological aspects of RNases and related enzymes in microorganisms.

# Chapter II

# Classification of Enzymes Attacking RNA

Enzymes of RNA metabolism may be classified into three categories: first, enzymes degrading RNA, such as ribonucleases (RNases) and polynucleotide phosphorylase; second, enzymes modifying RNA without degradation, such as methylating enzymes; third, enzymes of RNA synthesis, e.g., DNA-dependent RNA polymerase and RNA synthetase from RNA phage infected bacteria.

This monograph is concerned only with enzymes of degradation, and especially RNases, with reference to other related enzymes degrading poly- and oligo-ribonucleotides. Such enzymes found in microorganisms can be grouped as follows on the basis of their substrate specificities:

1. RNA-specific depolymerizing enzymes
   a) Ribonucleases
   b) Polynucleotide phosphorylase
2. Nonspecific nucleases
3. Phosphodiesterases
   a) Nonspecific phosphodiesterases
   b) Cyclic phosphodiesterase
4. Phosphomonoesterases
   a) Acid phosphatase
   b) Alkaline phosphatase.

RNases are usually defined as "phosphodiesterases" which can degrade RNA but do not hydrolyze either DNA or simple phosphodiesters such as bis-p-nitrophenyl phosphate. However, no strict definition of RNase can be adopted, as the modes of action of RNases are so various and certain "RNases" degrade oligodeoxyribonucleotides as well as RNA. Moreover from the biological point of view, it is rather inadequate to discuss RNases only in a narrow sense.

Accordingly, nucleolytic enzymes other than RNases will be treated in this monograph when necessary. In Chapter V especially, where the physiological roles of RNA degrading enzymes are considered, the properties and mode of action of polynucleotide phosphorylase (EC 2.7.7.8, Polynucleotide: orthophosphate nucleotidyltransferase) will be discussed together with those of RNases, because the former enzyme is known to play important roles in the metabolism of RNA, and the mutual relationship between various RNA degrading enzymes in cellular economy must be taken into consideration.

The enzymes which will be treated in this monograph are classified by their mode of action, reaction products, substrate specificities and modes of existence as shown in Table II-1. This classification is essentially similar to that suggested by LASKOWSKI. Such a classification is conventional, but no strict classification is possible, because RNases from different sources are more or less different although classified in the

Table II-1. *Classification of enzymes to be treated in this monograph*
*(Phosphorolytic enzymes are not included in this table)*

| Group | Sub-group | Notation in this monograph | Specification | Example | EC number |
|---|---|---|---|---|---|
| | | | Classification by the mode of action | | |
| I | 1 | tra | Degradation by nucleotidyl transfer | Ribonucleate nucleotido-2'-transferase (cyclizing) | 2.7.7.17 |
| | 2 | hyd | Degradation by hydrolysis | Ribonucleate (deoxyribonucleate) 3'-nucleotidohydrolase | 3.1.4.7 |
| II | 1 | endo | Endonucleolytic | Ribonucleate nucleotido-2'-transferase (cyclizing) | 2.7.7.17 |
| | 2 | exo | Exonucleolytic | Ribonucleate (deoxyribonucleate) | |
| | a | | from 3'-terminal | 5'-nucleotidohydrolase | 3.1.4.9 |
| | b | | from 5'-terminal | 3'-nucleotidohydrolase | 3.1.4.7 |
| | | | Classification by the reaction products | | |
| III | 1 | →pX | Enzymes producing 5'-nucleotides | Ribonucleate (deoxyribonucleate) 5'-nucleotidohydrolase | 3.1.4.9 |
| | 2 | | Enzymes producing 2',3'-cyclic phosphates | | |
| | a | >p | Enzymes producing 2',3'-cyclic phosphates as final products | Ribonucleate nucleotido-2'-transferase (cyclizing), *Azotobacter agilis* | 2.7.7.17 |
| | b | >p → Xp | Enzymes producing 3'-nucleotides as final products | Ribonucleate guaninenucleotido-2'-transferase (cyclizing) | 2.7.7.26 |
| | 3 | →Xp | Enzymes producing directly 3'-nucleotides | Ribonucleate (deoxyribonucleate) 3'-nucleotidohydrolase | 3.1.4.7. |

## Table II-1. (continued)

**Classification by the substrate specificity**

| | | | | | |
|---|---|---|---|---|---|
| IV | 1 | PDase | Nonspecific phosphodiesterase degrading polyribonucleotides | Orthophosphoric diester phosphohydrolase | 3.1.4.1 |
| | 2 | Nase | Nonspecific nuclease | Ribonucleate (deoxyribonucleate) 3'-nucleotidohydrolase | 3.1.4.7 |
| | 3 | RNase | Ribonuclease | | |
| | a | Non-s | non-base specific | Ribonucleate nucleotido-2'-transferase (cyclizing) | 2.7.7.17 |
| | b | Pyr-s | pyrimidine specific | Ribonucleate pyrimidinenucleotido-2'-transferase (cyclizing) | 2.7.7.16[1] |
| | c | G-s | guanine specific | Ribonucleate guaninenucleotido-2'-transferase (cyclizing) | 2.7.7.26 |
| | d | Pur-s | purine specific | Ribonucleate purinenucleotido-2'-transferase (cyclizing) | |

**Classification by the mode of existence**

| | | | |
|---|---|---|---|
| V | 1 | extra | extracellular |
| | 2 | intra | intracellular |
| VI | 1 | part | particulate |
| | 2 | sol | soluble |

[1] Not yet found in microorganisms.

same group. For example, most of the RNases classified as non-base specific RNases have their own relative base specificity. The characteristics of each enzyme will be discussed in Chapter IV.

For a general survey on RNases, special books and reviews should be consulted:

"The Enzymes" edited by D. BOYER, H. LARDY and K. MYRBÄCK, vol. 5, Academic Press, New York and London, 1961.

Chap. 1. "Phosphate Ester Cleavage (Survey)" by G. SCHMIDT and M. LASKOWSKI.

Chap. 6. "Phosphodiesterases" by H. G. KHORANA.

Chap. 8. "The Ribonucleases" by C. F. ANFINSEN and T. H. WHITE, JR.

"Les Nucléases" by M. PRIVAT DE GARILHE, Hermann, Paris, 1964.

"Procedures in Nucleic Acid Research" edited by G. L. CANTONI and D. R. DAVIES, Harper and Row, New York and London, 1966.

"Methods in Enzymology" edited by P. COLOWICK and N. O. KAPLAN, vol. 12, "Nucleic Acids", Academic Press, New York and London, 1967.

# Chapter III

# Distribution of RNases in Microorganisms
Microorganisms of which RNA-Degrading Enzymes have been Investigated

In the last few years there have been many reports on microbial RNA-degrading enzymes, namely RNases, phosphodiesterases and nucleases. Especially in Japan, for the purpose of producing 5'-IMP and 5'-GMP, on an industrial scale as food-seasoning agents, attempts to find an enzyme system capable of degrading RNA into 5'-ribonucleotides have been made on a wide range of microorganisms by screening procedures. Thus, for instance, KUNINAKA et al. studied the intracellular enzyme of *Penicillium sp.*, and OHMURA et al. the enzyme system in culture filtrates of *Strep-tomyces sp.* Although these enzyme systems in microorganisms are considered to be mixtures of ribonucleases, no details are known about their specificities except in a few strains.

The microorganisms of which RNA degradative enzymes have been studied are listed in Table III-1 with a brief note of the matters investigated and the reference in the literature. Enzymes that are fairly well characterized are treated in Chapter IV and their properties are summarized in Table IV-23.

Table III-1

I. Bacteria

| | | |
|---|---|---|
| *Thiobacillus thioparus* [a] | RNase, PDase [b]<br>[3'-Nucleotides (Py > Pu)] | [14—15] [c]<br>[16] |
| *Pseudomonas fluorescens* | Nuclease | [17] |
| *Pseudomonas aeruginosa* | (5'-Nucleotides)<br>Biosynthesis | [18]<br>[50] |
| *Azotobacter agilis* | Nuclease (Endonuclease)<br>(5'-Oligonucleotides)<br>RNase (Non-s) | [19]<br>[19]<br>[20—21] |
| *Agrobacterium tumefaciens* | Nuclease<br>RNase II | [23]<br>[22] |
| *Alcaligenes faecalis* | RNase-less strain | [24] |
| *Serratia marcescens* | Nuclease<br>Antigenicity of nuclease<br>Effect of nuclease on carcinoma<br>cells | [25], [28—29]<br>[26]<br><br>[27] |

[a] Microorganisms
[b] A brief note of matters investigated
[c] Reference

Table III-1 (continued)

| | | |
|---|---|---|
| *Escherichia coli* | Occurrence | [*9*], [*6*] |
| | RNase I | [*30—32*], [*35*] |
| | RNase II | [*33—34*], [*36*], [*38*] |
| | PDase, Depolymerase | [*46*], [*53*] |
| | Localization | [*39—43*], [*48*] |
| | RNA turn over in cells | [*44—45*], [*47*], [*51—52*] |
| | Biosynthesis | [*50*] |
| | Genetics | [*49*] |
| *Salmonella typhosa* | | |
| *Shigella flexneri* | | |
| *Mycobacterium avium* | RNase | [*9*] |
| *Micrococcus pyogenes* | | |
| *(Staphylococcus aureus)* | Endo, Exonuclease | [*9—10*], [*54—56*] |
| | Applied to TMV-RNA | [*57—61*] |
| | Purification, Properties | [*62—70*], [*72—74*] |
| | Review | [*71*] |
| *Micrococcus sodonensis* | PDase | [*75*] |
| *Staphylococcus epidermidis* | Nuclease | [*76*] |
| *Micrococcus lysodeikticus* | RNase | [*77*] |
| *Streptococcus hemolyticus* (Group A) | Nuclease (RNase) | [*3—4*], [*78*] |
| *Streptococcus pyogenes* | Nuclease | [*79*] |
| *Lactobacillus casei* | PDase (K ion-activated) | [*80*] |
| *Lactobacillus acidophilus* R-26 | Exonuclease (3'-Nucleotides) | [*81*] [*81*] |
| *Bacillus subtilis* | Extracellular, Intracellular RNase (H-strain) | [*9*], [*82—84*], [*90*] |
| | Properties | [*85*], [*88*], [*91*] |
| | Structure | [*86*], [*89*], [*96*] |
| | RNase, PDase (Marburg strain) | [*92*], [*95*] |
| | Exonuclease (SB-19 strain) | [*94*] |
| | (5'-Nucleotides) | [*18*] |
| | Review | [*87*] |
| | Miscellaneous | [*93*], [*93*][1], [*97*] |
| *Bacillus pumilus* | RNase (G-s) (3'-Nucleotides) | [*98*] [*99*] |
| *Bacillus brevis* | (5'-Nucleotides) | [*18*] |
| *Bacillus cereus* | RNase (Non-s) (3'-Nucleotides) | [*98*] [*99*] |
| *Bacillus licheniformis* | Occurrence in sporulation | [*100*] |
| *Bacillus amylozyma* | Nuclease | [*101—102*] |
| *Bacillus quercifolius* | Nuclease | [*101*], [*103*] |
| *Clostridium acetobutylicum* | (5'-Nucleotides) | [*104*] |
| *Clostridium perfringens* | RNase (PDase) | [*9*], [*105*] |
| *Clostridium histolyticum* | | |
| *Clostridium septicum* | | |
| *Clostridium tetani* | RNase | [*9*] |

## Table III-1 (continued)

| | | |
|---|---|---|
| *Asterococcus mycoides* | | |
| *(Mycoplasma mycoides)* | Nuclease | [*106*] |
| | (5′-APM, 3′-Nucleotides) | [*106*] |
| *Mycoplasma laidlawii* | | |
| *Mycoplasma bovigenitalium* | | |
| *Mycoplasma mycoides* var. | | |
| *mycoides* | | |
| *mycoides agalactiae* | | |
| *mycoides neurolyticum* | | |
| *mycoides mycoides* var. *capri* | | |
| *mycoides gallisepticum* | Nuclease (RNase) | [*107—108*] |
| | Distribution of RNases among | |
| | 200 strains of bacteria | [*53*] |

### II. Protozoa

| | | |
|---|---|---|
| *Tetrahymena* | Intracellular RNase | [*109*] |
| *Paramecium aurelia* | RNase | [*110*] |

### III. Streptomyces

| | | |
|---|---|---|
| *Actinomyces* (strain A) | RNase | [*5*] |
| *Streptomyces coelicolor* | | |
| *Streptomyces albogriseolus* | | |
| *Streptomyces aureus* | | |
| *Streptomyces gougerotii* | | |
| *Streptomyces griseoflavus* | | |
| *Streptomyces griseus* | | |
| *Streptomyces purpurescens* | | |
| *Streptomyces ruber* | | |
| *Streptomyces viridochromogenes* | (5′-Mononucleotides) | [*13*], [*18*] |
| *Streptomyces erythraeus* | RNase (G-s) | [*111—113*] |
| *Streptomyces albogriseolus* | RNase (G-s) | [*114—115*] |
| *Streptomycee aureus* | Nuclease (Endonuclease, PDase) | [*116—117*] |
| *Streptomyces aureoverticillatus* | RNase (G-s) | [*118*] |
| *Streptomyces hygroscopicus* | RNase (G-s) | [*119*] |
| *Actinomyces* | RNase (G-s) | [*120*] |
| *Streptomyces* sp. No. 41 | 5′-PDase (Exonuclease) | [*121—122*] |
| | 5′-Nucleotides | [*123*] |
| | 5′-PDase screening method | [*124*] |

### IV. Yeast

| | | |
|---|---|---|
| *Saccharomyces cerevisiae* | RNase (Non-s) | [*125—126*] |
| | 3′-Mononucleotides | |
| | (No 2′-, 3′-cyclic nucleotides) | [*125*] |
| *Saccharomyces cerevisiae* | | |
| *Candida utilis* | Localization | [*127*] |

Table III-1 (continued)

| | | |
|---|---|---|
| *Candida lipolytica* | | |
| *Candida claussenii* | | |
| *Candida albicans* | | |
| *Endomycopsis fibuliger* | | |
| *Hansenula anomala* | | |
| *Hansenula schneggii* | | |
| *Spolobolomyces pararoseus* | | |
| *Torulopsis candida* | | |
| *Torulopsis miso* | | |
| *Rhodotorula aurantiaca* | | |
| *Rhodotorula glutinis* | | |
| *Zygosaccharomyces tikumaensis* | Extracellular RNase | [99] |
| | (3'-Nucleotides) | [99] |
| | | |
| *Rhotorula glutinis* | Intracellular RNase, PDase | [128] |
| | (5'-, 3'-Nucleotides) | [128] |
| | Extracellular RNase | |
| | (3'-Nucleotides) | [128][1] |
| | | |
| *Endomyces* sp. | Intracellular RNase | |
| | (3'-Nucleotides) | [129] |

V. Fungi

| | | |
|---|---|---|
| *1) Fungi imperfecti* | | |
| *Aspergillus* sp. (Schimmelpilzen) | Occurrence | [2] |
| | | |
| *Aspergillus* sp. | | |
| *Penicillium* sp. | (5'-Nucleotides) | [132] |
| | | |
| *Aspergillus oryzae* | Nuclease | [7—8], [130] |
| | (3'-Nucleotides) | [131] |
| | (5'-Nucleotides) | [133] |
| | | |
| *Takadiastase* | Nuclease | [7] |
| | RNase $T_1$, $T_2$ (purification, properties) | [11] |
| | $T_1$, Method of purification | [140], [142—143] |
| | $T_1$ Properties | [134], [136—141] |
| | $T_1$ molecular weight, structure | [145—148] |
| | $T_1$ specificity applied to elucidate structure of s-RNA | [149—151], [151][1], [152] |
| | $T_2$ Purification, properties | [135], [156—159] |
| | Review | [162—163] |
| | Miscellaneous | [144], [153—155] |
| | | |
| *Aspergillus niger* | RNase | [160] |
| | | |
| *Aspergillus saitoi* | RNase (Non-s) | [161] |
| | | |
| *Aspergillus quercinus* | | |
| *Aspergillus elegans* | | |
| *Aspergillus fischeri* | | |
| *Aspergillus flavipes* | | |
| *Aspergillus melleus* | | |
| *Aspergillus nidulans* | | |
| *Aspergillus sulphureus* | | |
| *Aspergillus ustus* | (5'-Nucleotides)) | [18] |

Table III-1 (continued)

| | | |
|---|---|---|
| *Aspergillus fonsecaeus* | | |
| *Aspergillus flavus* | | |
| *Aspergillus quercinus* | (3'-Nucleotides) | [99] |
| *Penicillium citrinum* | PDase (Endo-, Exonuclease) | [12] |
| | (5'-Nucleotides) | [12], [164—165] |
| *Penicillium citrinum* | | |
| *Penicillium corymbiferum* | | |
| *Penicillium* baarnense var. | | |
| beyma | | |
| beyma asperum | | |
| beyma crustosum | | |
| beyma ehinulonalgiouense | (3'-Nucleotides) | [99] |
| *Penicillium commune* Thom | (3'-Nucleotides) | [166] |
| *Fusarium roseum* | | |
| *Fusarium solani* | | |
| *Gliomastix convoluta* | | |
| *Helminthosporium sigmoideum* var. | | |
| irregulare | | |
| *Verticillium niveostratosum* | (5'-Nucleotides) | [18], [167] |
| *Gloeosporium laeticolor* | | |
| *Pestalotia diospyri* | (3'-Nucleotides) | [99] |
| *Fusarium vasinfectum* | RNase | [169] |
| *Fusarium caucasicum* | RNase | |
| | No base specificity, | |
| | (2'-, 3'-cyclic nucleotides, | |
| | 3'-nucleotides) | [170] |
| *Sclerotium bataticota* | | |
| *Ascochyta sojaecola* | | |
| *Ascochyta phaseolorum* | (5'-Nucleotides) | [168] |
| *Diplodia* sp. | | |
| *Rhizoctonia solani* | | |
| *Phoma* sp. | | |
| *Septoria cucurbia* | | |
| *Phomopsis* sp. | | |
| *Phyllosticta sojaecola* | (5'-Nucleotides) | [171] |
| *Acrocylindrium* sp. | RNase, PDase | |
| | RNase I, II (G-s) | [170], [172] |
| | (5'-Nucleotides) | [18] |
| *Scopulariopsis brevicaulis* var. | | |
| glabra | | |
| *Trichothecium roseum* | RNase (Non-s) | |
| | (2'-, 3'-Nucleotides) | [170] |
| *Stysanus stemonitis* | RNase (G-s) | [170] |
| 2) *Ascomycetes* | | |
| *Neurospora crassa* | RNase | [173], [178], [178][1] |
| | (5'-Nucleotides) | [18] |
| | Nuclease | [174—175], [177] |
| | RNase N$_1$ | [176] |
| | RNase N$_1$, N$_2$, N$_3$ | [176][1] |

Table III-1 (continued)

| | | |
|---|---|---|
| *Neurospora sitophila* | (5′-Nucleotides) | [*18*] |
| *Monascus ruber* | | |
| *Monascus anka* | | |
| *Monascus major* | | |
| *Monascus fuliginosus* | | |
| *Monascus rubiginosus* | | |
| *Monascus pilosus* | (3′-Nucleotides) | [*99*] |
| *Monascus purpureus* | RNase (3′-, 5′-Nucleotides) | [*179*] |
| *Monascus pilosus* | RNase (Non-s) | [*98*] |
| *Anixiella reticulispora* | | |
| *Chaetomidium japonicum* | | |
| *Glomerella cingulata* | | |
| *Ophiobolus miyabeanus* | | |
| *Sordaria fimicola* | | |
| *Ophiostoma ulmi* | (5′-Nucleotides) | [*18*] |
| *Gibberella fujikuroi* | | |
| *Sclerotinia fuckeliana* | (3′-Nucleotides) | [*99*] |
| *3) Basidiomycetes* | | |
| *Merulius demosticus* | | |
| *Poria vaporaria* | | |
| *Irpex lacteus* | | |
| *Polystictus hirsutus* | | |
| *Polystictus versicolor* | (3′-Nucleotides) | [*99*] |
| *Polystictus hirsutus* | RNase (Non-s) | [*170*] |
| *Lenzites tenuis* | (3′-Nucleotides) | [*99*] |
| | RNase (Non-s) | [*98*] |
| *Agaricus campestris* | (3′-Nucleotides) | [*180*] |
| *Ustilago zeae* | RNase (G-s) | [*181*] |
| *Ustilago sphaerogena* | RNase (G-s) | [*182*], [*187*] |
| | RNase (Pur-s) | [*187*] |
| | RNase (Non-s) | [*187*] |
| *Corticium sasakii* | | |
| *Pellicularia filamentosa* | (5′-Nucleotides) | [*183*] |
| *Pellicularia* sp. | (5′-Nucleotides) | [*184*], [*184*][1] |
| *4) Phycomycetes* | | |
| *Rhizopus chinensis* | | |
| *Rhizopus niveus* | (3′-Nucleotides) | [*99*] |
| *Rhizopus niveus* | RNase (2′-, 3′-Cyclicnucleotides) | [*185*] |
| *Phytophthora infestans* | RNA degradation | [*186*] |
| *Absidia coerulea* | | |
| *Absidia spinosa* | | |
| *Absidia butleri* | | |
| *Blackeslea circinans* | | |
| *Cunninghamella echinulata* | | |
| *Helicostylum piriforme* | | |
| *Mucor pusillus* | | |
| *Mucor genevensis* | | |
| *Phycomyces blakesleeanus* | (3′-Nucleotides) | [*99*] |
| *Absidia butleri* | RNase (Non-s) | [*170*] |
| *Mucor genevensis* | RNase (G-s) | [*98*] |

Table III-1 (continued)

1. NOGUCHI, J.: Biochem. Z. **147**, 255 (1924).
2. OTANI, H.: Acta Schol. Med. Univ. Imp. Kioto **17**, 323 (1935).
3. McCARTY, M.: J. Exp. Med. **88**, 181 (1948).
4. TILLETT, W. S., S. SHERRY, and L. R. CHRISTENSEN: Proc. Soc. Exp. Biol. Med. **68**, 184 (1948).
5. MUGGLETON, P. W., and M. WEBB: Biochim. et Biophys. Acta **9**, 343 (1952).
6. PARDEE, A. B., and R. E. KUNKEE: J. Biol. Chem. **199**, 9 (1952).
   —, and I. WILLIAMS: Arch. Biochem. Biophys. **40**, 222 (1952).
7. KUNINAKA, A.: J. Agr. Chem. Soc. Japan (Nippon Nogei-kagaku Kaishi) **28**, 282 (1954).
8. SARUNO, R.: Bull. Agr. Chem. Soc. Japan **20**, 57 (1956).
9. MASUI, M., Y. HONDA, K. HIKITA, T. IWATA, I-HUNG PAN, and I. TAKI: Osaka City Med. J. **7**, 141 (1956).
10. CUNNINGHAM, L., B. W. CATLIN, and M. PRIVAT DE GARILHE: J. Am. Chem. Soc. **78** 4642 (1956).
11. SATO, K., and F. EGAMI: J. Biochem. (Tokyo) **44**, 753 (1957).
12. KUNINAKA, A., and K. SAKAGUCHI: Japan Pat., 36-5287 (1961).
13. OHMURA, E., K. OGATA, Y. SUGINO, S. IGARASI, M. YONEDA, Y. NAKAO, and I. SUHARA: Japan Pat., 36-16630 (1961).
14. OSTROWSKI, W.: Experientia **17**, 398 (1961).
15. —, and Z. WALCZAK: Acta Biochim. Polon. **8**, 345 (1961).
16. WALCZAK, Z., and W. OSTROWSKI: Acta Biochim. Polon. **11**, 241 (1964).
17. KYUNE, M. F., and I. E. PAVLOVSKAYA: Uchenye Zapiski, Kazan. Gosudarst. Univ. im V. I. Ul'yanova-Lenina **124**, 153 (1964).
18. OGATA, K., Y. NAKAO, S. IGARASI, E. OHMURA, Y. SUGINO, M. YONEDA, and I. SUHARA: Agr. Biol. Chem. **27**, 110 (1963).
19. STEVENS, A., and R. J. HILMOE: J. Biol. Chem. **235**, 3016, 3023 (1960).
20. SHIIO, I., and S. SHIMIZU: J. Japan. Biochem. Soc. (Seikagaku) **34**, 406 (1962).
21. —, K. ISHII, and S. SHIMIZU: J. Biochem. (Tokyo) **59**, 363 (1966).
22. REDDI, K. K.: Proc. Nat. Acad. Sci. US **56**, 1207 (1966).
23. BEAUD, G.: Compt. rend. **257**, 2752 (1963).
24. NATORI, S., T. HORIUCHI, and D. MIZUNO: Biochim. et Biophys. Acta **134**, 337 (1967).
25. EAVES, G. N., and C. D. JEFFRIES: J. Bacteriol. **85**, 273, 1194 (1963).
26. NUZHINA, A. M., and T. M. TRET'YAYA: Uchenye Zapiski, Kazan. Gosudarst Univ. im V.I. Ul'yanova-Lenina **124**, 131 (1964).
27. KYUNE, M. F., and J. E. PAVOLOVSKAYA: Uchenye Zapiski, Kazan. Gosudarst Univ. im V.I. Ul'yanova-Lenina **124**, 148 (1964).
28. MESSINOVA, O. V.: Uchenye Zapiski, Kazan. Gosudarst Univ. im V.I. Ul'yanova-Lenina **124**, 122 (1964).
29. LESCHINSKAYA, I. B., and Z. F. BOGAUTDINOV: Mikrobiologiya **32**, 412 (1963).
30. ELSON, D.: Biochim. et Biophys. Acta **27**, 216 (1958).
31. — Biochim. et Biophys. Acta **36**, 372 (1959).
32. SPAHR, P. F., and B. R. HOLLINGWORTH: J. Biol. Chem. **236**, 823 (1961).
33. —, and D. SCHLESSINGER: J. Biol. Chem. **238**, PC2251 (1963).
34. — J. Biol. Chem. **239**, 3716 (1964).
35. — Procedures Nucleic Acid Res. **1966**, 64.
36. SINGER, M. F.: Procedures Nucleic Acid Res. **1966**, 192.
37. —, and G. TOLBERT: Science **145**, 593 (1964).
38. — — Biochemistry **4**, 1319 (1965).
39. NEU, H. C., and L. A. HEPPEL: Biochem. Biophys. Res. Commun. **17**, 215 (1964).
40. — — Proc. Nat. Acad. Sci. US **51**, 1267 (1964).
41. — — J. Biol. Chem. **239**, 3893 (1964).
42. — — J. Biol. Chem. **240**, 3685 (1965).
43. ANRAKU, Y., and D. MIZUNO: Biochem. Biophys. Res. Commun. **18**, 462 (1965).
44. MARUYAMA, H., and D. MIZUNO: Biochim. et Biophys. Acta **108**, 593 (1965).
45. ANDERSON, J. H., and C. E. CARTER: Biochemistry **4**, 1102 (1965).

Table III-1 (continued)

46. LEHMAN, I. R.: J. Biol. Chem. 235, 1479 (1960); Procedures Nucleic Acid Res. 203 (1966).
47. FUTAI, M., Y. ANRAKU, and D. MIZUNO: Biochim. et Biophys. Acta 119, 373 (1966).
48. NOSSAL, N. G., and L. A. HEPPEL: J. Biol. Chem. 241, 3055 (1966).
49. GESTELAND, R. F.: J. Mol. Biol. 16, 67 (1966).
50. OLEINIK, I. I., S. I. BESBORODOVA, V. V. KACALUKHA, and G. P. SMIRNOVA: Mikrobiologiya 35, 220 (1966).
51. NOSE, K., D. MIZUNO, and H. OZEKI: Biochim. et Biophys. Acta 119, 636 (1966).
52. NAKAJIMA, K., and J. KAWAMATA: Biken J. 9, 115 (1966).
53. WADE, H. E., and H. K. ROBINSON: Biochem. J. 101, 467 (1966).
54. PRIVAT DE GARILHE, M., L. CUNNINGHAM, ULLA-RIITTA LAURILA, and M. LASKOWSKI: J. Biol. Chem. 224, 751 (1957).
55. CUNNINGHAM, L.: J. Am. Chem. Soc. 80, 2546 (1958).
56. — Ann. N. Y. Acad. Sci. 81, 788 (1959).
57. REDDI, K. K.: Nature 182, 1308 (1958).
58. — Biochim. et Biophys. Acta 36, 132 (1959).
59. — Nature 188, 60 (1960).
60. — Biochim. et Biophys. Acta 47, 47 (1961).
61. RUSHIZKY, G. W., C. A. KNIYHT, W. K. ROBERTS, and C. A. DEKKER: Biochim. et Biophys. Acta 55, 674 (1962).
62. McEVOY, C. S.: Marquette Med. Rev. 31, 137 (1965).
63. POCHON, F., and M. PRIVAT DE GARILHE: Bull. soc. chim. biol. 42, 795 (1960).
64. ALEXANDER, M., L. A. HEPPEL, and J. HURWITZ: J. Biol. Chem. 236, 3014 (1961).
65. ANFINSEN, C. B., M. K. RUMLEY, and H. TANIUCHI: Acta Chem. Scand. 17, S270 (1963).
66. OHSAKA, A., J. MUKAI, and M. LASKOWSKI, SR.: J. Biol. Chem. 239, 3498 (1964).
67. MUKAI, J., A. OHSAKA, C. McEVOY, and M. LASKOWSKI, SR.: Biochem. Biophys. Res. Communn. 18, 136 (1965).
68 SULKOWSKI, E., and M. LASKOWSKI, SR.: J. Biol. Chem. 241, 4386 (1966).
69. COTTON, F. A., E. E. HAZEN, JR., and D. C. RICHARDSON: J. Biol. Chem. 241, 4389 (1966).
70. TANIUCHI, H., and C. B. ANFINSEN: J. Biol. Chem. 241, 4366 (1966).
71. HEINS, J. N., H. TANIUCHI, and C. B. ANFINSEN: Procedures Nucleic Acid Res. 1966, 79.
72. CHESBRO, W. R., D. STUART, and J. J. BURKE: Biochem. Biophys. Ros. Commun. 23, 783 (1966).
73. BARKER, G. R., and J. G. PAVLIK: Biochem. J. 98, 4p (1966).
74. HEINS, J. N., J. R. SURIANO, H. TANIUCHI, and C. B. ANFINSEN: J. Biol. Chem. 242, 1016 (1967).
75. BERRY, S. A., and J. N. CAMPBELL: Biochim. et Biophys. Acta 132, 78, 84 (1967).
76. BELYAEVA, M. L., D. V. YUSOPOVA, and E. G. DEDYUKHINA: Uchenye Zapiski, Kazan. Gosudarst. Univ. im V.I. Ul'yanova-Lenina 124, 49 (1964).
77. BARKER, G. R., and M. CANNON: Biochem. J. 75, 8p (1960).
78. McCLEAN, D.: J. Pathol. Bacteriol. 53, 13 (1941).
79. PAVIOVA, I. M.: Mikrobiol. Zhur., Akad. Nauk Ukr. R.S.R., Inst. Mikrobiol. im D.K. Zabolotnogo 27, 12 (1965).
80. KEIR, H. M., R. H. MATHOG, and C. E. CARTOR: Biochemistry (U.S.S.R.) 3, 1188 (1964).
81. FIERS, W., and H. G. KHORANA: J. Biol. Chem. 238, 2780, 2789 (1963); Arch. intern. physiol. et biochem. 71, 299 (1963).
82. NISHIMURA, S., and M. NOMURA: Biochim. et Biophys. Acta 30, 430 (1958).
83. — — J. Biochem. (Tokyo) 46, 161 (1959).
84. —, and B. MARUO: Biochim. et Biophys. Acta 40, 355 (1960).
85. — Biochim. et Biophys. Acta 45, 15 (1960).
86. —, and H. OZAWA: Biochim. et Biophys. Acta 55, 421 (1962).
87. — Procedures Nucleic Acid Res. 1966, 56.
88. RUSHIZKY, G. W., A. E. GRECO, R. W. HARTLEY, JR., and H. A. SOBER: Biochemistry 2, 787 (1963).

Table III-1 (continued)

89. HARTLEY, JR., R. W., G. W. RUSHIZKY, A. E. GRECO, and H. A. SOBER: Biochemistry 2, 794 (1963).
90. RUSHIZKY, G. W., A. E. GRECO, R. W. HARTLEY, JR., and H. A. SOBER: Biochem. Biophys. Res. Commun. 10, 311 (1963).
91. WHITFIELD, P. R., and H. WITZEL: Biochim. et Biophys. Acta 72, 362 (1963).
92. NAKAI, N., Z. MINAMI, T. YAMAZAKI, and A. TSUGITA: J. Biochem. (Tokyo) 57, 96 (1965).
93. SMEATON, J. R., W. H. ELLIOTT, and G. COLEMAN: Biochem. Biophys. Res. Commun. 18, 36 (1965).
93¹. COLEMAN, G., and W. H. ELLIOTT: Biochem. J. 95, 699 (1965).
94. KERR, I. M., E. A. PRATT, and I. R. LEHMAN: Biochem. Biophys. Res. Commun. 20, 154 (1965).
95. TANIGUCHI, K., and A. TSUGITA: J. Biochem. (Tokyo) 60, 372 (1966).
96. LEES, C. W., and R. W. HARTLEY, JR.: Biochemistry 5, 3951 (1966).
97. COWGILL, R. W.: Biochim. et Biophys. Acta 120, 189 (1966).
98. RUSHIZKY, G. W., A. E. GRECO, R. W. HARTLEY, JR., and H. A. SOBER: J. Biol. Chem. 239, 2165 (1964).
99. NAKAO, Y., and K. OGATA: Agr. Biol. Chem. 27, 116 (1963).
100. LEITZMANN, C., and R. W. BERNLOHR: J. Bacteriol. 89, 1506 (1965).
101. BENING, G. P., O. V. MESSINOVA, and S. V. MALITTSEVA: Uchenye Zapiski, Kazan. Gosudarst Univ. im V.I. Ul'yanova-Lenina 124, 101 (1964).
102. KYUNE, M. F.: Uchenye Zapiski, Kazan. Gosudarst Univ. im V.I. Ul'yanova-Lenina 124, 136 (1964).
103. ZELENKOVA, N. P.: Uchenye Zapiski, Kazan. Gosudarst Univ. im V.I. Ul'yanova-Lenina 124, 157 (1964).
104. KATAGIRI, H., and T. SUGIMORI: Japan Pat., 38-20587 (1963).
105. MUKHIN, I. V.: Uchenye Zapiski, Kazan. Gosudarst Univ. im V.I. Ul'yanova-Lenina 124, 88 (1964).
106. PLACKETT, P.: Biochim. et Biophys. Acta 26, 664 (1957).
107. RAZIN, S., A. KNYSZYNSKI, and Y. LIFSHITZ: J. Gen. Microbiol. 36, 323 (1964).
108. POLLACK, J. D., S. RAZIN, and R. C. CLEVERDON: J. Bacteriol. 90, 617 (1965).
109. ROTH, J. S.: Ann. N. Y. Acad. Sci. 81, 611 (1959).
110. GROSS, G., B. SKOCZYLAS, and W. TURSKI: Acta Protozool. 4, 59 (1966).
111. TANAKA, K.: J. Biochem. (Tokyo) 50, 62 (1961).
112. —, and G. L. CANTONI: Biochim. et Biophys. Acta 72, 641 (1963).
113. — Procedures Nucleic Acid Res. 1966, 14.
114. YONEDA, M.: 13th Symposium of Enzyme Chemistry in Japan, 143 (1961).
115. — J. Biochem. (Tokyo) 55, 469 (1964).
116. — 18th Symposium of Enzyme Chemistry in Japan, 38 (1962).
117. — J. Biochem. (Tokyo) 55, 475, 481 (1964).
118. TATARSKAYA, R. I., A. I. KORENYAKO, and N. M. ABROSIMOVA-AMEL'YANCHIK: Prinkl. Biokhim. i Mikrobiol. 2, 151 (1966).
119. TA-CHUEH TSO: Sheng Wu Hua Hsueh Yu Sheng Wu Wu Li Hsueh Pao 5, 17 (1965).
120. TATARSKAYA, R. I.: Doklady Akad. Nauk S.S.S.R. 157, 725 (1964).
121. SUGIMOTO, H., T. IWASA, J. ISHIYAMA, and T. YOKOTSUKA: J. Agr. Chem. Soc. Japan (Nippon Nogei-kagaku Kaishi) 37, 677 (1963).
122. — —, and T. YOKOTSUKA: J. Agr. Chem. Soc. Japan (Nippon Nogei-kagaku Kaishi) 38, 135, 567 (1964).
123. YOKOTSUKA, T.: Japan Pat., 40-3466 (1965).
124. SUGIMOTO, H., T. IWASA, J. ISHIYAMA, and T. YOKOTSUKA: J. Agr. Chem. Soc. Japan (Nippon Nogei-kagaku Kaishi) 36, 277 (1962).
125. OHTAKA, Y., K. UCHIDA, and T. SAKAI: J. Biochem. (Tokyo) 54, 322 (1963).
126. RANNER, J., and R. S. MORGAN: Biochim. et Biophys. Acta 76, 652 (1963).
127. SCHLENK, F., and J. L. DAINKO: J. Bacteriol. 89, 428 (1965).
128. NAKAO, Y., I. NOGAMI, and K. OGATA: Agr. Biol. Chem. 27, 507 (1963).
128¹. —, and K. OGATA: Agr. Biol. Chem. 27, 499 (1963).

Table III-1 (continued)

129. HATTORI, Y., and S. NAKAMURA: J. Japan. Biochem. Soc. (Seikagaku) **38**, 563 (1966).
130. KUNINAKA, A.: J. Agr. Chem. Soc. Japan (Nippon Nogei-kagaku Kaishi) **29**, 52, 797, 801 (1955).
131. — J. Gen. Appl. Microbiol. **3**, 55 (1957).
132. FUJISHIMA, T., I. UCHIDA, and H. YOSHINO: Japan Pat., 41-16538 (1966).
133. ANDO, T.: Biochim. et Biophys. Acta **114**, 158 (1966).
134. SATO-ASANO, K.: J. Biochem. (Tokyo) **46**, 31 (1959).
135. TADA, M., K. ASANO, and F. EYAMI: J. Biochem. (Tokyo) **46**, 757 (1959).
136. SATO-ASANO, K., and Y. FUJII: J. Biochem. (Tokyo) **47**, 608 (1960).
137. — J. Biochem. (Tokyo) **48**, 284 (1960).
138. WHITFIELD, P. R., and H. WITZEL: Biochim. et Biophys. Acta **72**, 338 (1963).
139. UCHIDA, T., and F. EGAMI: J. Biochem. (Tokyo) **57**, 742 (1965).
140. — J. Biochem. (Tokyo) **57**, 547 (1965).
141. RUSHIZKY, G. W., and H. A. SOBER: J. Biol. Chem. **237**, 834, 2883 (1962).
142. CHUNG, H., and S. MANDELES: Biochim. et. Biophys. Acta **92**, 403 (1964).
143. MINATO, S., T. TAGAWA, and K. NAKANISHI: J. Biochem. (Tokyo) **59**, 443 (1966).
144. HIRAMARU, M., T. UCHIDA, and F. EGAMI: Anal. Biochem. **17**, 135 (1966).
145. TAKAHASHI, K.: J. Biochem. (Tokyo) **49**, 1 (1961).
146. — J. Biochem. (Tokyo) **51**, 95 (1962).
147. — J. Biochem. (Tokyo) **52**, 72 (1962).
148. — J. Biol. Chem. **240**, PC4117 (1965).
149. McCULLY, K. S., and G. L. CANTONI: Biochim. et Biophys. Acta **51**, 190 (1961).
150. STAEHELIN, M.: Biochim. et Biophys. Acta **87**, 493 (1964).
151. HOLLEY, R. W., G. A. EVERETT, J. T. MADISON, and A. ZAMIR: J. Biol. Chem. **240**, 2122 (1965).
151[1]. —, J. APGAR, G. A. EVERETT, J. T. MADISON, M. MARQUISEE, S. H. MERRILL, J. R. PENSWICK, and A. ZAMIR: Science **147**, 1462 (1965).
152. DÜTTING, D., and H. G. ZACHAU: Z. physiol. Chem. **336**, 132 (1964).
153. SHORTMAN, K.: Biochim. et Biophys. Acta **55**, 88 (1962).
154. IRIE, M.: J. Biochem. (Toky ) **56**, 495 (1964).
155. SATO, S., and F. EGAMI: Biochem. Z. **342**, 437 (1965).
156. UCHIDA, T.: J. Biochem. (Tokyo) **60**, 115 (1966).
157. RUSHIZKY, G. W., and H. A. SOBER: J. Biol. Chem. **238**, 371 (1963).
158. TADA, M.: J. Japan. Biochem. Soc. (Seikagaku) **38**, 662 (1966).
159. SATO, S., T. UCHIDA, and F. EGAMI: Arch. Biochem. Biophys. **115**, 48 (1966).
160. YANAGITA, Y., and F. KOGANE: J. Japan. Biochem. Soc. (Seikagaku) **34**, 406 (1962).
161. IRIE, M.: J. Japan. Biochem. Soc. (Seikagaku) **38**, 563 (1966).
162. UCHIDA, T., and F. EGAMI: Procedures Nucleic Acid Res. **1966**, 46.
163. EGAMI, F., K. TAKAHASHI, and T. UCHIDA: Progress in Nucleic Acid Res. and Mol. Biol. **3**, 59 (1964).
164. KUNINAKA, A., S. OTSUKA, Y. KOBAYASHI, and K. SAKAGUCHI: Bull. Agr. Chem. Soc. Japan **23**, 239 (1959).
165. —, M. KIBI, H. YOSHINO, and K. SAKAGUCHI: Agr. Biol. Chem. **25**, 693 (1961).
166. SUGIMOTO, H., T. IWASA, and J. ISHIYAMA: J. Agr. Chem. Soc. Japan (Nippon Nogei-kagaku Kaishi) **36**, 690 (1962).
167. OHMURA, E.: Japan Pat., 37-4544 (1962).
168. TONE, H., H. SASAKI, Y. SAYAMA, T. ISHIKURA, and N. MIYAJI: Japan Pat., 40-13800 (1965).
169. SAMPATHNARAYANAN, A., and E. R. B. SHANMUGASUNDARAM: Phytopathol. Z. **57**, 79 (1966).
170. OHMURA, E., I. SUHARA, F. KUSABA, and Y. NAKAO: 6th Int. Congress of Biochem. New York, Abstract 1-149 (1964).
171. TONE, H., Y. SAYAMA, T. ISHIKURA, and N. MIYAJI: Japan Pat., 40-15959 (1965).
172. SUHARA, I., F. KUSABA, and E. OHMURA: 16th Symposium of Enzyme Chemistry in Japan 115 (1964).

Table III-1 (continued)

173. SUSKIND, S. R., and D. M. BONNER: Biochim. et Biophys. Acta **43**, 173 (1960).
174. LINN, S., and I. R. LEHMAN: J. Biol. Chem. **240**, 1287, 1294 (1965).
175. SOMBERG, E. W., and F. F. DAVIS: Biochim. et Biophys. Acta **108**, 137 (1965).
176. TAKAI, N., T. UCHIDA, and F. EGAMI: Biochim. et Biophys. Acta **128**, 218 (1966).
176[1]. — — — J. Japan Biochem. Soc. (Seikagaku) **39**, 285 (1967).
177. LINN, S., and I. R. LEHMAN: J. Biol. Chem. **241**, 2694 (1966).
178. TOHE, A., K. HASUNUMA, and T. ISHIKAWA: 38th Annual Meeting Genetics Soc. Japan, 23 (1966).
178[1]. MURAYAMA, H., K. ISONO, and T. ISHIKAWA: 38th Annual Meeting Genetics Soc. Japan, 23 (1966).
179. SARUNO, R., H. TAKAHIRA, and M. FUJIMOTO: J. Fermentation Technol. (Hakko Kogaku Zasshi) **42**, 475 (1964).
180. CUIGNIEZ, J., L. DEHENNIN, J. STOCKX und L. VANDENDRIESSCHE: Naturwissenschaften **52**, 187 (1965).
181. YANAGIDA, M., T. UCHIDA, and F. EGAMI: J. Agr. Chem. Soc. Japan (Nippon Nogei-kagaku Kaishi) **38**, 531 (1964).
182. GLITZ, D. G., and C. A. DEKKER: Biochemistry **3**, 1391 (1964).
183. UCHI, M., and Y. HASEGAWA: Japan Pat., 41-16120 (1966).
184. HASEGAWA, Y., T. NAKAI, Y. FUJIMURA, Y. KANEKO, and S. DOI: J. Agr. Chem. Soc. Japan (Nippon Nogei-kagaku Kaishi) **38**, 461 (1964).
184[1]. FUJIMURA, Y., Y. HASEGAWA, Y. KANEKO, and S. DOI: J. Agr. Chem. Soc. Japan (Nippon Nogei-kagaku Kaishi) **38**, 467 (1964).
185. KOGA, K., J. MUKAI, and S. AKUNE: J. Fac. Agr., Kyushu Unuv. **13**, 711 (1966).
186. PAGE, O. T.: Phytophathology **55**, 259 (1965).
187. ARIMA, T., T. UCHIDA, and F. EGAMI: 7th International Biochem. Congress in Tokyo (1967), Abstract. B-20.

Consideration of the specificities of RNases, indicates the following general characteristics:

1. In microorganisms, RNases which specifically attack the bonds next to pyrimidine nucleotides have not yet been found.

2. Almost all the species investigated have proved to have intracellular RNase activity and they generally have no specificity for nucleotides.

3. RNases which are specific for guanine nucleotide are extracellular enzymes released from bacteriophyta and eumycophyta, for instance *Aspergillus* and *Neurospora*. Among extracellular RNases, of course, nonspecific enzymes have been found. However, recently RNases were found which were specific for purine nucleotides besides others in the culture filtrate of *Ustilago sphaerogena*.

It is very interesting to trace the changes in specificity during the evolution of organisms in relation to the transition of structure of RNase protein. So far, all RNases have been found to have a rather small molecular size and nonspecific ones have a molecular weight of about 30,000 to 40,000 while base specific RNases have one of 11,000 to 13,000. Moreover, properties such as heat stability and optimum pH are strikingly similar in all kinds of RNases found in a wide range of organisms from microbes to mammals. This suggests that RNases may be good materials for anylsis of protein from the viewpoint of comparative biochemistry.

However, RNases have unfortunately been studied in only a few species and therefore it is difficult at present to discuss these enzymes in relation to evolution.

It is expected that in future RNases will be studied in species at various stages of evolution and their chemical natures will be compared.

# Chapter IV

# Chemical Studies on Microbial RNases

## A. Introduction

Chemical studies of RNases are significant from both chemical and biological points of view.

RNases are generally enzyme proteins of relatively small molecular size. They have no prosthetic group and require no special cofactor for enzymatic activity. Most of them are fairly heat-stable. Because of these properties, RNases are the most suitable enzymes for use in elucidating the relationship between protein structure and enzymatic activity. This is also the reason why pancreatic RNase (RNase I-A) has been fully investigated by STEIN and MOORE, ANFINSEN, and many others, and recently also by D. HARKER. Moreover, microbial RNases have the added advantage that they can be produced on a large scale by tank culture of microorganisms.

The primary structures of proteins such as hemoglobin from various animals have been elucidated from an interest in comparative biochemistry. However, proteins such as hemoglobins are present only in a relatively limited number of species of animals[1]. RNases, on the other hand, are widely distributed in animals, plants, and microorganisms and, as mentioned above, they have a relatively simple molecular structure. So RNases may be regarded as the best proteins for studies on comparative biochemistry and evolutionary biochemistry at the molecular level.

In spite of these outstanding characteristics of RNases, there have been few reports of chemical studies on microbial RNases.

Here the chemical properties of more or less well purified microbial RNases will be described. As the purification procedures are essentially the same as those of other proteins, they will not be described here in detail: for details, the original papers should be consulted. However, it should be mentioned that the heat-stability and small molecular size of various RNases facilitate their separation from other proteins.

From the chemical point of view, the microbial RNase which has been studied most is RNase $T_1$ of *Aspergillus oryzae*. So we will first discuss the RNases of *Aspergillus oryzae* in detail and then the ribonucleases from other sources will be mentioned for comparison.

In this chapter only RNases in the strict sense will be mentioned, i.e., enzymes degrading RNA, but hydrolyzing neither DNA (nor denatured DNA), nor bis-*p*-nitrophenyl phosphate.

---

[1] Hemoglobin is also present in root nodules of some plants.

## B. RNases of Aspergillus oryzae

Although a long time ago IWANOFF (1903), NOGUCHI (1924), and OTANI (1935) reported the degradation of RNA by *Aspergillus oryzae* or by Takadiastase (a commercial product from this mold), the existence of RNases in *Aspergillus oryzae* was only recently described by KUNINAKA (1955) and SARUNO (1956). In this connection, the critical survey by KUNINAKA (1957) of the enzymatic degradation of yeast RNA by *Asp. oryzae* is a valuable paper.

It was reported that a nuclease in Takadiastase was as thermostable as the familiar pancreatic RNase (RNase I, or more exactly RNase I-A). This finding led one of the authors (F. E.) to consider that it would be interesting to compare the chemical nature of the former enzyme with that of the latter, which has been fully investigated by STEIN and MOORE, ANFINSEN and many others. Thus, studies on RNases in Takadiastase were undertaken in 1956 by EGAMI in collaboration with Miss K. SATO (later Mrs. K. ASANO).

As it was found in preliminary experiments that Takadiastase contained at least three enzymes which attacked RNA, the enzymatic properties of the major component were studied first. A study of the substrate specificity of the enzyme led to the unexpected discovery that, unlike pancreatic RNase, it split the phosphodiester bonds of 3'-guanylic acid in RNA. One of the minor components was found to split the phosphodiester bonds of 3'-adenylic acid in RNA preferentially. EGAMI and SATO named the guanylic-acid specific RNase and the RNase preferentially attacking adenylic acid bonds RNase T$_1$ and RNase T$_2$, respectively (SATO and EGAMI, 1957). As will be described later, it was found that RNase T$_2$ degraded RNA practically completely to four 3'-mononucleotides without showing absolute base specificity. So according to the Commission on Enzymes of the International Union of Biochemistry, these enzymes may be classified as follows:

| Individual name | EC number | Systematic name | Trivial name |
|---|---|---|---|
| RNase T$_1$ | 2.7.7.26 | Ribonucleate 2'-guaninenucleo-tido-2'-transferase (cyclizing) | Guanyloribonuclease |
| RNase T$_2$ | 2.7.7.17 | Ribonucleate nucleotido-2'-trans-ferase (cyclizing) | Ribonuclease |

### 1. RNase T$_1$

Ribonucleate guaninenucleotido-2'-transferase (cyclizing), *Aspergillus oryzae;* tra, endo, > p → Xp, G-s, extra

*Purificatior of RNase T$_1$*: Since the partial purification of RNase T$_1$ by SATO and EGAMI, several modifications of the procedure have been reported. TAKAHASHI (1962) first obtained the enzyme as a homogeneous protein in the form of fine spherular crystals. UCHIDA (1965) simplified the purification procedure, as will be described below.

RUSHIZKY and SOBER (1962) purified RNase T$_1$ by somewhat different procedures; the product appears to be nearly as pure as that prepared by TAKAHASHI's or UCHIDA's procedures. MINATO et al. of Sankyo Co. Ltd (Tokyo, Japan) developed a large-scale

2*

purification procedure for the commercial preparation of RNase $T_1$ (1963) and they reported that they could obtain RNase $T_1$ as relatively large crystals (1966).

Although RNase $T_1$ is available commercially, it may be prepared without difficulty in the laboratory by UCHIDA's procedure (1965, 1966): RNase $T_1$ in the crude extract of Takadiastase powder is adsorbed on DEAE-cellulose equilibrated with 0.005 $M$ $Na_2HPO_4$ by batchwise treatment. From the DEAE-cellulose carrying adsorbed RNase $T_1$ and $T_2$, RNase $T_2$ is first eluted with 0.1 $M$ NaCl in 0.01 $M$ $Na_2HPO_4$, and then RNase $T_1$ is eluted with 0.35 $M$ NaCl in 0.01 $M$ $Na_2HPO_4$. The RNase $T_1$ fraction is adjusted to pH 1.5 to 1.8 with HCl, and the precipitate formed is removed by centrifugation. The clear, yellowish-brown supernatant is heated at 80° for 2 min and then cooled quickly. This heating procedure serves to inactivate other contaminating enzymes. To the heat-treated RNase $T_1$ fraction (pH 1.5) is added Japanese acid clay, which adsorbs the enzyme. The cake of acid clay is eluted with citrate buffer, pH 6.3. The eluate is brought to 70% saturation of $(NH_4)_2SO_4$ and the suspension is centrifuged. The supernatant fluid is adjusted to pH 4.2 and brought to 100% saturation of $(NH_4)_2SO_4$. The precipitate is collected by centrifugation, dissolved in a small amount of citrate buffer, pH 6.3, and dialyzed against running tap water. The dialyzed RNase $T_1$ is subjected to DEAE-cellulose column chromatography. The fractions of high specific activity (500 to 700 fold purification over the crude extract) are combined and precipitated at 70% saturation of $(NH_4)_2SO_4$ at pH 4.0. The precipitates are dissolved in a small amount of 0.1 $M$ citrate buffer at pH 6.3 and dialyzed in the cold room against 50 volumes of $(NH_4)_2SO_4$ solution of 60% saturation, at pH 4.0. The enzyme gradually precipitates out as colorless, spherular crystals. The enzyme crystals are dissolved in a small amount of 0.1 $M$ citrate buffer at pH 6.3 and stored in a deep-freezer after dialysis against distilled water.

*Properties of RNase $T_1$* (EGAMI et al., 1964).

1. pH Optimum and enzyme activity

RNase $T_1$ is most active at about pH 7.5. At pHs of about 5.5 and 8.0, 50% of the maximum activity remains. RUSHIZKY and SOBER reported that the optimal pH is 7.3 and that 50% of the optimal activity remains at pH 5.2 and 7.9.

The optimum pH of RNase $T_1$ for the hydrolysis of guanosine nucleoside 2',3'-cyclic phosphate is in the same pH region as that for RNA digestion (RUSHIZKY and SOBER, 1962).

2. Stability

RNase $T_1$ is as stable a protein as RNase I-A. At neutrality and at acidic pH values it is quite stable. However, at pH values above pH 9, it is somewhat unstable.

The enzyme is fairly resistant to heat: no loss of activity was observed after heating it in solution at 100 °C for 10 min (pH 6.0).

3. Factors affecting activity

The effects of various substances other than nucleotides on RNase $T_1$ are summarized in Table IV-1.

As shown in Table IV-1, $Mg^{2+}$ and $Ca^{2+}$ slightly inhibit RNase $T_1$. $Zn^{2+}$ causes strong inhibition. The apparent stimulatory action of EDTA (ethylenediaminetetraacetate) might be explained as due to elimination of contaminating metallic ions.

RUSHIZKY and SOBER reported that their preparation of RNase T$_1$ was not activated by EDTA (RUSHIZKY and SOBER, 1962).

The apparent activation by a low concentration of histidine (1 m$M$) might be attributed to the elimination of inhibitory metal ions such as Zn. Since $p$-chloromercuribenzoate and diisopropyl phosphorofluoridate have no effect on RNase T$_1$, it appears that this enzyme is neither a SH enzyme nor a "serine enzyme". The former conclusion is consistent with the primary structure of RNase T$_1$.

Table IV-1. *Inhibitors and activators*
*of RNase T$_1$*

| Reagents | Final concentration (—log $M$) | Activity remaining (%) RNase T$_1$ |
|---|---|---|
| NaCl | 0 | 75 |
| | 1 | 100—115 |
| NaF | 1 | 93—100 |
| NaN$_3$ | 2 | 95—100 |
| Na$_2$S | 2 | 10 |
| AgNO$_3$ | 3 | 0 |
| MgCl$_2$ | 1 | 60 |
| CaCl$_2$ | 2 | 70—75 |
| HgCl$_2$ | 3 | 10 |
| MnSO$_4$ | 2 | 45 |
| ZnSO$_4$ | 3 | 0 |
| CuSO$_4$ | 3 | 20—50 |
| FeSO$_4$ | 2 | 20 |
| ICH$_2$COOH | 2 | 100 |
| BrCH$_2$COOH | 4 | 100 |
| DFP | 6 | 100 |
| Histidine | 2 | 82 |
| | 3 | 150 |
| EDTA | 2 | 125—150 |

4. Physical and chemical properties

The physical and chemical characteristics of RNase T$_1$ are summarized in Table IV-2. RNase T$_1$ has a molecular weight of 11,000 which is less than that of RNase I-A. It is a simple protein with alanine as the N-terminal, and threonine as the C-terminal amino acid. RNase T$_1$ is an acidic protein. In this respect it is quite different from pancreatic RNase (RNase I-A), which is a basic protein.

RUSHIZKY and SOBER (1962) reported some of the properties of RNase T$_1$, and these are in good agreement with our findings.

Heparin, which strongly inhibits RNase I-A, does not inhibit RNase T$_1$. This may be explained by the consideration that heparin, which is a polyanionic macromolecule, binds basic proteins, such as RNase I-A, but not acidic proteins such as RNase T$_1$.

According to K. SHORTMAN (1962), a natural RNase I-A inhibitor does not affect RNase T$_1$. Quite recently, UOZUMI et al. (1967) found that a natural inhibitor of a non-specific nuclease obtained from *Asp. oryzae* does not inhibit RNase $T_1$. It is remarkable that natural inhibitors such as this are highly specific.

## 5. Chemical nature (TAKAHASHI, 1966)

*Amino acid composition.* The amino acid composition of RNase $T_1$ is given in Table IV-3 in comparison with that of RNase I-A. RNase $T_1$ is characterized by a low content of basic amino acids. Thus, there is only one residue of arginine and one

Table IV-2. *Physical and chemical properties of RNase $T_1$*

|  | RNase $T_1$ | RNase I-A |
|---|---|---|
| Molecular weight |  |  |
| Sedimentation diffusion | 11000 | 12700 |
| Sedimentation equilibrium | — | 14000 |
| Amino acid analysis | 11085[a] | 13683 |
| $S_{20,w}$ | 1.62 S | 1.85 S ($S_{25}$) |
| $D_{20,w}$ | $12.0 \times 10^{-2}$ cm² sec⁻¹ | $13.6 \times 10^{-2}$ cm² sec⁻¹ ($D_{25}$) |
| $f/f_0$ | 1.21 | — |
| Electrophoretic mobility ($\mu = 0.1$) | $-2.82 \times 10^{-4}$ cm² V⁻¹ sec⁻¹ (pH 7.0) | — |
|  | $-0.71_3 \times 10^{-4}$ cm² V⁻¹ sec⁻¹ (pH 4.0) | — |
|  | $+0.25_1 \times 10^{-4}$ cm² V⁻¹ sec⁻¹ (pH 2.5) | — |
| Isoelectric point | pH 2.9 | pH 7.8 |
| Absorption maximum | 278 mμ | 277.5 mμ |
| Absorption minimum | 251—252 mμ | — |
| OD max/OD min | $3.0_1$ | — |
| $OD_{max\ 10m}^{0.1\%}$ | $1.9_1$ | $0.71_6$ |
| $[\alpha]_D$ | $-24°$ | $-71.7°$ |
| Nitrogen content | 16.5% | 16.5% |
| Sugar content | $0 \sim\ < 0.5\%$ | — |
| N-terminal amino acid | Ala | Lys |
| C-terminal amino acid | Thr | Val |
| pH optimum for RNA digestion | 7.4 | 7.0—7.5 |

[a] Based on the complete amino acid sequence.

of lysine. It also has a high content of glycine (12 residues), one tryptophan residue and no methionine. These results indicate that RNase $T_1$ is rather different in primary structure from pancreatic RNase.

The remarkable high acidity of the enzyme can be attributed to the low content of cationic groups (6 residues) and relatively high content of anionic groups (15 residues). Despite the lower content of lysine and arginine, the enzyme contains three histidine residues per molecule. In this it is comparable to RNase I-A. This fact is noteworthy, for histidine residues might be involved in the catalytic function of RNase $T_1$ as in that of RNase I-A. RNase $T_1$ contains no free SH group like RNase I-A. So RNase $T_1$ may be regarded as having two cystine residues.

*Primary structure.* The primary structure of RNase T₁ has been elucidated by TAKAHASHI (1965). The terminal groups were analyzed by Sanger's DNFB method and Edman's PTH method for the amino terminal and by Akabori's hydrazinolysis method and the carboxypeptidase method for the carboxyl terminal.

Although native RNase T₁ is resistant to trypsin and chymotrypsin, it is easily digested after heat denaturation or performic acid oxidation. As the primary step in the elucidation of the primary structure of RNase T₁, RNase T₁ which had been oxidized with performate was digested with trypsin and chymotrypsin and the resulting peptides were separated and analyzed.

A tentative structure for RNase T₁ was deduced by combining the information obtained on the amino acid sequences of these peptides. Peptic digestion of the heat-

Table IV-3. *Amino acid composition of RNase T₁*

| Amino acid | RNase T₁ | RNase I-A |
|---|---|---|
| Aspartic acid | 15 | 15 |
| Threonine | 6 | 10 |
| Serine | 15 | 15 |
| Glutamic acid | 9 | 12 |
| Proline | 4 | 4 |
| Glycine | 12 | 3 |
| Alanine | 7 | 12 |
| Half-cystine | 4 | 8 |
| Valine | 8 | 9 |
| Methionine | 0 | 4 |
| Isoleucine | 2 | 3 |
| Leucine | 3 | 2 |
| Tyrosine | 9 | 6 |
| Phenylalanine | 4 | 3 |
| Lysine | 1 | 10 |
| Histidine | 3 | 4 |
| Arginine | 1 | 4 |
| Tryptophan | 1 | 0 |
| Amide ammonia | (12) | (17) |
| Total | 104 | 124 |

denatured RNase T₁ and papain digestion and Nagarse (subtilisin) digestion of the native enzyme were carried out to confirm the results and determine the location of the disulphide bonds.

Thus the complete amino acid sequence was determined. Further, two peptides each containing one cystine residue and a peptide containing two cystine residues were isolated from the Nagarse digestion products, and were identified (Chart IV-1).

From these results the complete primary structure of RNase T₁ was determined to be as shown in Chart IV-2.

The most remarkable characteristic of the structure of RNase T₁ is the position of the two disulphide bonds. As they are located in the vicinity of the N- and C-terminals, the total structure resembles a rope with its two ends bound.

This primary structure of RNase T₁ may be compared with that of RNase I-A (Chart IV-3) which was elucidated by STEIN and MOORE, and ANFINSEN et al. If we assume the existence of a catalytic site and a specific binding site at the active center, comparison of the two should indicate a similarity in the catalytic site and a difference in the specific binding site. Although no definite conclusion can be reached, certain speculations can be made.

<pre>
  1   2   3   4              5   6   7   8    9
Ala-Cys-Asp-Tyr          Thr-Cys-Gly-Ser-AspNH₂
     |                            |
    Cys-Tyr                     Cys-Thr
    11  12                      103 104

    S-XXXVIIIα                  S-XXVα

 1   2 ┌ 3   4             5   6   7    8      9       10 ┐ 11
Ala-Cys-Asp-Tyr          Thr-Cys-G ly-Ser-AspNH₂-Cys-Tyr
                                          |
         AspNH₂-Phe-Val-Glu-Cys-Thr
          99     100   101   102  103  104

              S-XXXVIII β
</pre>

Chart IV-1. Cystine containing peptides isolated from a 5 h subtilisin digest of RNase T₁

One of the disulphide bonds in RNase T₁ forms a ring structure consisting of 9 amino acid residues with alanine and tyrosine adjacent to the disulphide bonds. There is a similar ring structure consisting of 8 residues in RNase I-A.

It remains to be shown whether this apparent similarity has any significance in relation to the activities of the enzymes. However, there seems to be little similarity between RNase I-A and RNase T₁. This is to be expected as the two enzymes are

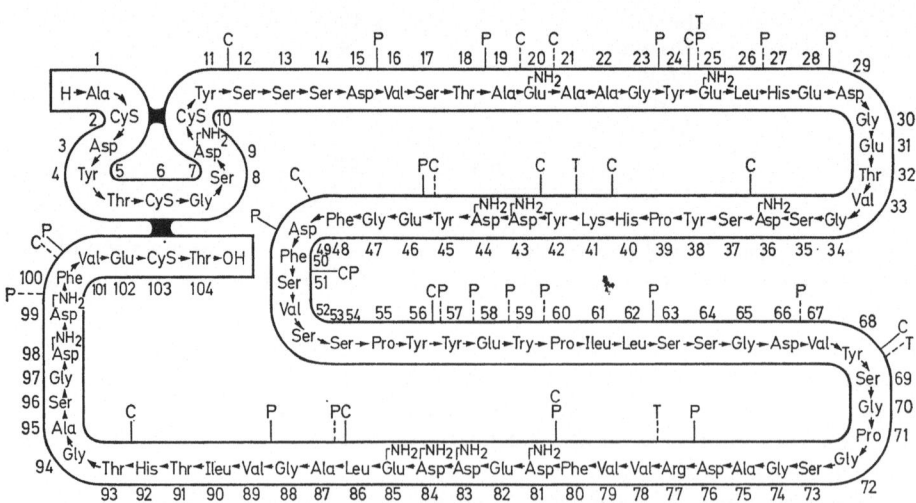

Chart IV-2. Amino acid sequence of RNase T₁ [The points of hydrolysis by trypsin and chymotrypsin in the performic acid-oxidized protein and by pepsin in the heat-denatured protein are marked by T, C and P respectively. The solid lines represent extensive or rapid hydrolysis, and the dashed lines, incomplete or slower hydrolysis]

derived from quite different sources, one from a mammalian gland and the other from a mold, and they have different substrate specificities.

As will be discussed later, one or two histidine residues seem to be essential for the activities of both RNase T₁ and RNase I-A. The disulphide bonds are essential for maintenance of the active structures of the two enzymes.

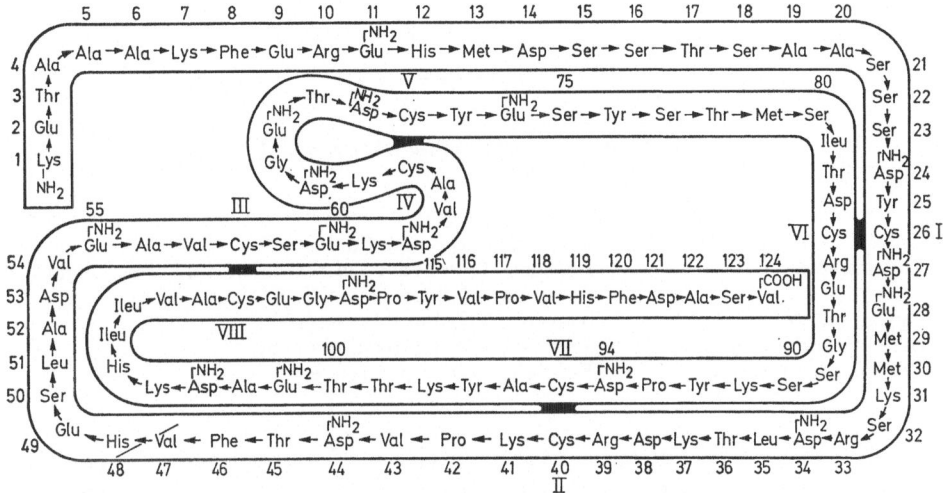

Chart IV-3. Amino acid sequence of RNase I-A

*Conformation.* J. TANAKA and his coworkers are carrying out extensive optical investigations on RNase T₁. They have investigated the ultraviolet absorption spectra of the enzyme over a wide pH range. Between pH 7.2 and 9.9, the absorption spectrum remains essentially unchanged with an absorption maximum at 278 mμ, and over a wide range on the alkaline side (pH 10.85) another almost constant absorption spectrum is observed with an absorption maximum at 291 mμ. This suggests that the tyrosine residues in the enzyme are embedded in the helical structure of the enzyme molecule and around pH 10.20 the helical structure disappears. The transition from a helix to a random coil is also observed by measuring CD or ORD. The helical content of the native enzyme is estimate to be about 35% (M. YAMAMOTO and J. TANAKA, 1967).

*Relation between structure and function.* The relationship between structure and function has been investigated by chemical and enzymatic modifications of the enzyme protein.

1. Action of proteases (TAKAHASHI, 1962; SAIGUSA et al., 1961)

The effects of various proteases on RNase T₁ are summarized in Table IV-4, RNase T₁ is resistant to the action of trypsin, chymotrypsin, and leucine aminopeptidase, but sensitive to pepsin. RNase T₁ is sensitive to carboxypeptidase A, suggesting that a sequence near the C-terminal participates in some way in the enzymatic activity. However, the terminal threonine can be removed without loss of activity. The effects of various proteases on RNase $T_1$ are somewhat similar to those of proteases on RNase I-A.

Partial hydrolysis of RNase $T_1$ by Nagarse is now being investigated, to see if "RNase $T_1$" with a lower molecular weight can be obtained and also to see if interesting phenomena similar to that observed by RICHARDS in the case of RNase I-A can be seen.

## 2. Chemical modifications

a) Photooxidation (YAMAGATA et al., 1962). RNase $T_1$ was inactivated by methylene blue-catalysed photo-oxidation. Photooxidative decomposition of one of the three

Table IV-4. *Enzymatic inactivation of RNase $T_1$*

| Enzyme | RNase $T_1$ |
|---|---|
| Trypsin | Resistant, fully active (in 0—6 $M$ urea) |
| Chymotrypsin | Resistant, fully active (in the absence of urea); slightly sensitive, 12% inactivation (in 6 $M$ urea, 20 h) |
| Pepsin | Sensitive, 70% inactivation (22 h) |
| Carboxypeptidase A | Sensitive, 65% inactivation (44 h) |
| Leucine aminopeptidase | Resistant, almost fully active |

histidine residues in RNase $T_1$ resulted in almost 90% inactivation (Fig. IV-1), suggesting the importance of at least one of the histidine residues. Besides histidine, tryptophan was also decomposed, though to a lesser extent. Thus the role of tryptophan in enzyme activity cannot be ruled out.

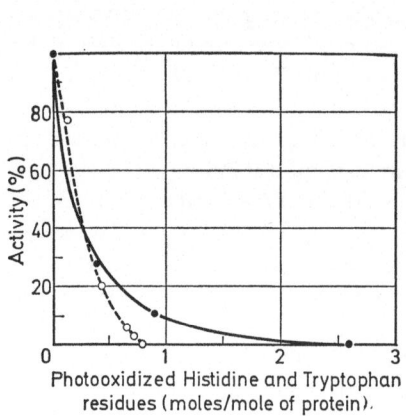

Fig. IV-1

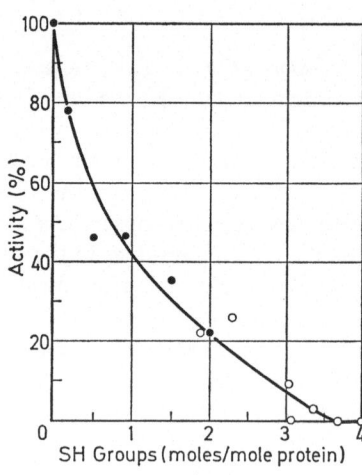

Fig. IV-2

Fig. IV-1. Relationship between the photo-oxidation of histidine and tryptophan residues and the changes in the enzyme activity of RNase $T_1$ [●——●, histidine; and o - - - o, tryptophan]

Fig. IV-2. Relationship between the degree of disulphide cleavage and the changes in the enzyme activity of RNase $T_1$ [Reduction was carried out in the presence (o) or absence (●) of 7.2 $M$ urea]

b) Deamination (Shiobara et al., 1962). As mentioned earlier, RNase T₁ is a very acidic protein, containing only two free amino groups, the $\alpha$-amino group of the N-terminal alanine and the $\varepsilon$-amino group of one lysine residue in the molecule. The two amino groups in RNase T₁ are deaminated with nitrous acid at 2 °C. Deamination of the $\alpha$-amino group of the amino terminal alanine occurs rapidly and has no effect on the enzyme activity. Deamination of the $\varepsilon$-amino group of the sole lysine residue occurs slowly together with modifications of some of the tyrosine residues. Even after complete deamination of both amino groups, appreciable activity remains. These results indicate that neither of the two amino groups are directly involved in the expression of the enzyme activity. Recently, this was confirmed by finding that when RNase T₁ was trinitrophenylated at both amino groups it retained full activity (Kassai et al., 1965). It is rather remarkable that a protein molecule deprived of free amino groups is still enzymatically active.

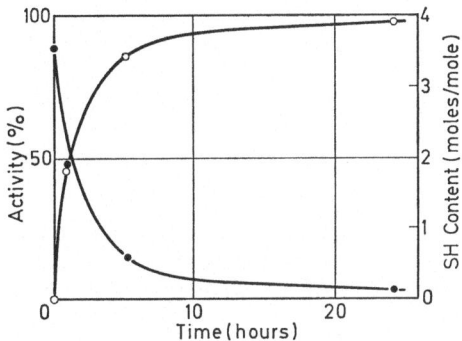

Fig. IV-3. Regeneration of enzymatic activity and disappearance of free SH groups on oxidation of reduced RNase T₁ in air [The solution of reduced RNase T₁ (1.7 mg per ml) in 0.05 $M$ tris HCl (pH 8.0) was stored at room temperature. o — o, enzymatic activity; and ● — ●, free SH groups]

c) Reductive cleavage (Yamagata et al., 1962; Kasai, 1965). Reductive cleavage of the disulphide bonds by thioglycolate or mercaptoethanol inactivated RNase T₁ (Fig. IV-2).

d) Reoxidation (Yamagata et al., 1962; Kasai, 1965). However, the enzyme activity was regenerated by oxidation in air (Fig. IV-3). These results demonstrate the importance of disulphide bonds for the maintenance of the enzymatically active structure.

e) Rotatory dispersion. The rotatory dispersion was estimated with native and reduced RNase T₁ in 8 $M$ urea solution, and with native and reduced-reoxidized RNase T₁ in the absence of urea. The reduced-reoxidized RNase T₁ may be regarded as identical with the native enzyme (Fig. IV-4, Table IV-5).

f) Carboxymethylation. RNase T₁ was inactivated by carboxymethylation with bromoacetate in weak acidic medium. In the presence of 2'-guanylic acid, which competitively inhibits RNase T₁, the enzyme was scarcely inactivated by bromoacetate (Sato and Egami, 1965). By analogy with RNase I-A, the working hypothesis was adopted that 2'-guanylic acid binds the histidine residue in the active centre and protects the enzyme from inactivation by bromoacetate. In accordance with the base specificity of RNase $T_1$, 2'-cytidcyclic acid did not protect the enzyme from inactivation by bromoacetate. The finding that 5'-guanylic acid scarcely protected RNase $T_1$ from inactivation by bromoacetate suggests the importance of the position of the phosphate group in nucleotides in their binding to the enzyme (Fig. IV-5).

Quite recently TAKAHASHI, MOORE, and STEIN (1967) have found that the residue attacked by this reagent is not a histidine residue but a glutamyl residue and,

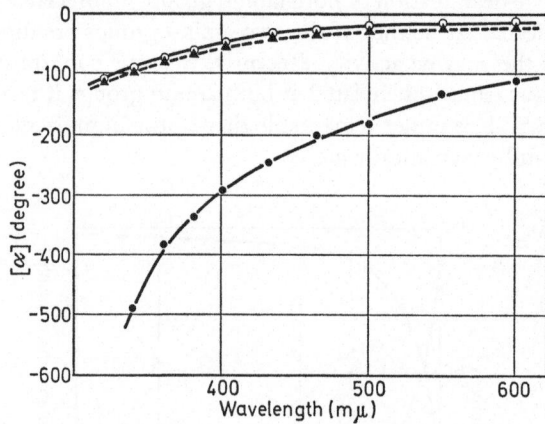

Fig. IV-4. Optical rotatory dispersion curves of RNase $T_1$ [o — o, native RNase $T_1$ (5.9 mg/ml); ● — ●, RNase $T_1$ in 8 $M$ urea (8.0 mg/ml) in the presence or absence of β-mercaptoethanol; and ▲ - - - ▲, reoxidized RNase $T_1$ (7.2 mg/ml)]

Table IV-5. *Comparison of some properties of native, reduced and reducedreoxidized RNase $T_1$*

| Property | Native RNase $T_1$ | Reduced RNase $T_1$ | Reduced-reoxidized RNase $T_1$ |
|---|---|---|---|
| Enzyme activity (%) | 100 | <0.3 | 100 |
| Free SH (moles/mole) | 0 | ~3.3 | 0 |
| — $[\alpha]_{400}$ | 51.0° | 293° [a] | 54.3° |
| $\lambda c$ (mμ) | 243 | 212 [a] | 244 |
| UV absorption max (mμ) | 278—279 | 276—277 | 278—279 |

[a] In 8 M urea
$\lambda c$, rotatory dispersion constant

using RNase $T_1$ which had been inactivated by reaction for 5 h with iodoacetate (1.5%) at pH 5.5 and 37°, they isolated a hexapeptide, Tyr-"Glu"-Trp-Pro-Ile-Leu,
                                                                    57      58      59   60   61   62
in which "Glu" was shown to be esterified as follows:

$$.....-Tyr-CO-CH_2-NH-Trp-.....$$
$$|$$
$$CH_2$$
$$|$$
$$CH_2$$
$$|$$
$$COOCH_2COO^-$$

This is a very important discovery not only for elucidation of the nature of the active center of the enzyme, but also from the viewpoint of protein chemistry in general.

*Substrate specificity.* RNase T₁ and RNase T₂ can hydrolyze neither DNA nor bis-*p*-nitrophenyl phosphate. For this reason they are classified as RNases rather than as nonspecific phosphodiesterases.

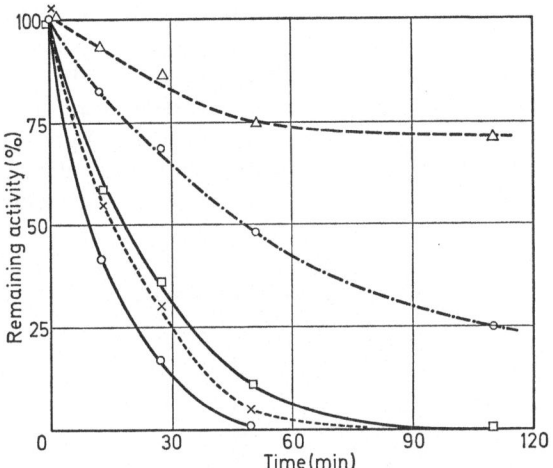

Fig. IV-5. Effect of nucleotides on the reaction of RNase T₁ with bromoacetate in 0.02 *M* acetate buffer (pH 5.5), at 27 °C. [Final concentration of RNase T₁, 7.5 μ*M* and that of bromoacetate, 0.25 *M*. o — o, no addition; △ - - - △, in the presence of 2′-GMP (0.28 m*M*); o —.— o, in the presence of 3′-GMP (0.28 m*M*); □ — □, in the presence of 5′-GMP (0.28 m*M*); and ✕ ..... ✕, in the presence of 2′-CMP (0.52 m*M*)]

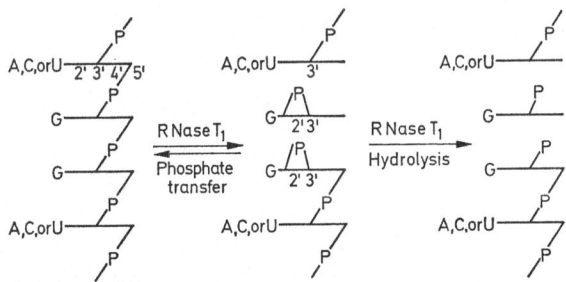

Chart IV-4. Mechanism of action of RNase T₁ indicating its substrate specificity

RNase T₁ splits the internucleotide bonds between 3′-guanylic acid groups and the 5′-hydroxyl groups of the adjacent nucleotides with the intermediary formation of guanosine 2′,3′-cyclic phosphate. Hence, RNase T₁ may be regarded as a guanylic acid-specific endoribonuclease. The mechanism of this reaction (Chart IV-4) was deduced from the following experimental evidence: (1) RNase T₁ digests yeast RNA and produces only one mononucleotide, 3′-guanylic acid; (2) the terminal residues of the oligonucleotides produced are exclusively guanylic acid; and (3) guanosine-2′,3′-

cyclic phosphate is obtained in good yield as an intermediate of the reaction and is further hydrolyzed to 3'-guanylic acid (SATO and EGAMI, 1957; SATO-ASANO, 1959).

As RNase $T_1$ does not split the secondary phosphate ester bonds of adenosine 3'-phosphate, its specificity is considered to depend on the substituents of the purine ring. This consideration led us to study the effect of RNase $T_1$ on deaminated RNA. RNase $T_1$ was found to digest deaminated RNA, hydrolyzing the secondary phosphate ester-bonds of both inosine 3'-phosphate and xanthosine 3'-phosphate, though the latter reaction was far slower. In both cases, nucleoside 2',3'-cyclic phosphates were formed as intermediates (SATO-ASANO and EGAMI, 1960).

These results suggest that the essential requirement in the structure of the base for susceptibility to RNase $T_1$ may be the oxo (or hydroxy) group at the 6-position of the purine base. Other substituents in the purine base more or less affect the susceptibility to RNase $T_1$. The 2-oxo group in xanthylic acid residues decreases the susceptibility.

Table IV-6. *Relative rates of splitting of different substrates by RNase $T_1$*

| Compound | Relative rate |
|---|---|
| GpCp | 1100 |
| GpC | 800 |
| GpA | 550 |
| GpG | 450 |
| GpU | 250 |
| IpC | 150 |
| XpC | 10 |
| Glyoxal-GpC | 5 |
| G-cyclic-p | 2 |

Further studies on the base specificity have been carried out by WHITFELD and WITZEL with several dinucleotides and related compounds as substrates (Table IV-6) (1963).

In the course of studies on RNase $T_1$ digestion of s-RNA, CANTONI, HOLLEY, STAEHELIN and their coworkers investigated the susceptibility of methylated guanylate bonds to RNase $T_1$. Methylated guanylate bonds in s-RNA were found to be less susceptible to the enzyme than guanylate bonds. Results of studies on the substrate specificity are summarized in Table IV-7. The results further support the absolute requirement of the oxo group at the 6-position of the purine base. The loss of a proton at N-1 of the purine base greatly decreases the susceptibility.

HIRAMARU et al. found that RNA methylated with dimethyl sulfate is quite resistant to RNase $T_1$ at the optimum pH (7.4) of the enzyme. They considered that this might be explained by the consideration that the methylation products, 7-methylguanine residues, in RNA were dissociated at the optimum pH, as shown in Chart IV-5.

However, they found recently that methylated RNA is resistant to RNase $T_1$ even at pH 4.5, where the oxo-group at position 6 is not dissociated and N-1 is protonated as in guanylate. So the possibility that N-7 in purine base participates in the susceptibility to RNase $T_1$ is not excluded.

Besides those mentioned above, few studies have been carried out on the action of RNase T$_1$ on modified RNA. Riboapyrimidinic acid, resistant to RNase I-A, is hydrolyzed as well as RNA by RNase T$_1$ (TAKEMURA and MIYAZAKI, 1959). According to AZEGAMI and IWAI (1964), RNA trinitrophenylated at the 2-amino group of the guanylyl residues in RNA is resistant to RNase T$_1$.

Table IV-7. *Base specificity of RNase T$_1$*

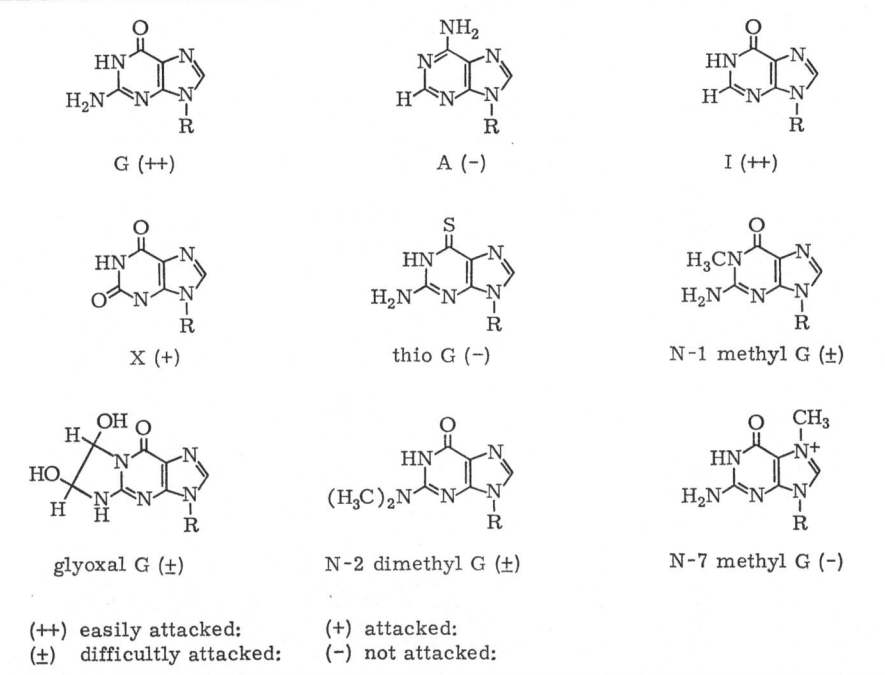

| | | |
|---|---|---|
| G (++) | A (−) | I (++) |
| X (+) | thio G (−) | N-1 methyl G (±) |
| glyoxal G (±) | N-2 dimethyl G (±) | N-7 methyl G (−) |

(++) easily attacked:  (+) attacked:
(±) difficultly attacked:  (−) not attacked:

It should be added here that IRIE reported that poly A, poly C and poly U could be digested by a higher concentration of RNase T$_1$ (1965). This may be explained either as due to contamination of his preparation with other enzymes, such as RNase

I

Chart IV-5

T$_2$, or to a special behavior of homopolymers to RNase T$_1$. With natural RNA as substrate, adenylyl, cytidilyl and uridilyl bonds could not be cleaved at all.

As shown in the above diagram, the first step (phosphate transfer) in the enzymatic digestion is reversible, guanylic acid polymers are produced from guanosine

$2',3'$-cyclic phosphate by RNase $T_1$ in concentrated substrate solution (SATO-ASANO and EGAMI, 1958; SATO-ASANO and EGAMI, 1960; HAYASHI and EGAMI, 1963). These oligoguanylates are completely digested to monomers by RNase $T_1$ in dilute solution. On the other hand, oligoguanylates which have been synthesized chemically (by MICHELSON) are only partially hydrolyzed by RNase $T_1$ (HAYASHI and EGAMI, 1963). This may be explained by assuming that the latter contains, besides $3',5'$-phosphodiester bonds, $2',5'$-phosphodiester bonds which are resistant to RNase $T_1$. This consideration was further supported by the finding that chemically synthesized polyguanylic acid was only partly hydrolyzed by spleen phosphodiesterase. Moreover, the core resistant to RNase $T_1$ is resistant to spleen phosphodiesterase, and vice versa (KURIYAMA et al., 1964). As it is firmly established that spleen phosphodiesterase

Table IV-8. *Inhibition of ribonuclease $T_1$ by various nucleotides and dissociation constants of enzyme-inhibitor complexes at pH 7.0*

| Compound | Degree of inhibition [a] (%) | $K_i \times 10^4$ (M) | Compound | Degree of inhibition [a] (%) | $K_i \times 10^4$ (M) |
|---|---|---|---|---|---|
| None | 0 | | | | |
| Guanosine-2'-P | 93[1] | 0.34 | Inosine-5'-P | 17[1] | |
| Guanosine-3'-P | 78[1] | 0.66 | Cytidine-2'-P | 60[2] | 3.40 |
| Guanosine-5'-P | 70[1] | 1.10 | Cytidine-3'-P | 50[2] | 6.80 |
| Adenosine-2'-P | 77[1] | | Cytidine-5'-P | 26[2] | |
| Adenosine-5'-P | 58[1] | | Uridine-2'(3')-P | 20[2] | |
| Inosine-2'-P | 40[1] | | Uridine-5'-P | 6[2] | |
| Inosine-3'-P | 35[1] | | Inorganic P | 26[3] | |

[a] Degree of inhibition $= 100 \times \left(1 - \dfrac{\text{initial velocity in the presence of inhibitor}}{\text{initial velocity in the absence of inhibitor}}\right)$.

The reaction mixture consisted of 1 ml of substrate (40 mg of G-cyclic-p in 10 ml of water), 1 ml of 0.1 $M$ NaCl and 1 ml of inhibitor or water.

The reaction was carried out at 35 °C using a pH-stat. The experimental conditions for $K_i$ measurement were the same as above except for the substrate concentration (from 0.7 to 7 m$M$). The final concentrations of the inhibitors were as follows: [1] 0.85 m$M$, [2] 1.1 m$M$ and [3] 2 m$M$.

hydrolyzes $3',5'$-phosphodiester bonds but not $2',5'$-phosphodiester bonds, it may be concluded that RNase $T_1$ is also inactive on $2',5'$-phosphodiester bonds.

This is supported by the observation that a mixture of dinucleoside monophosphates, such as a mixture of Gp-(2', 5')-N and Gp-(3', 5')-N, is only 50% cleaved by the enzyme.

Concerning studies on specificity, is should be added that guanosine $3',5'$-cyclic phosphate, unlike $2',3'$-cyclic phosphate, is quite resistant to RNase $T_1$ (SATO et al., unpublished data). It is now quite certain that RNase $T_1$ hydrolyzes the phosphodiester bonds of the guanylyl residues in RNA. Are all these bonds hydrolyzed irrespectively of the nature of the neighboring groups and other conditions? All these bonds in natural RNA, including highly polymerized RNA prepared by the method of CRESTFELD et al. and sRNA, are completely hydrolyzed. However, it should be mentioned that certain polynucleotides rich in guanylyl residues are sometimes fairly resistant to RNase $T_1$. This is the case with the fraction rich in guanylic acid from the RNase I-A core of yeast RNA (ITAGAKI et al., 1965). This may be due to the aggregation of polynucleotide molecules (ISHIKURA, 1962).

*Observations on the binding of RNase $T_1$ with 2'-guanylate and other related compounds.* M. IRIE has studied the competitive inhibition of RNase $T_1$ by various nucleotides (1964). The results are summarized in Table IV-8.

It is consistent with the specificity of the enzyme that guanylates have higher affinity to the enzymes than other nucleotides. Among guanylates, 2'-guanylate has the highest affinity. This has been confirmed by the spectroscopic observations of SATO and EGAMI on the interaction of nucleotides and RNase $T_1$ (1965). They have found that one RNase $T_1$ molecule binds one 2'-guanylate molecule, that is to say, the RNase $T_1$ molecule has one active site responsible for the specific binding.

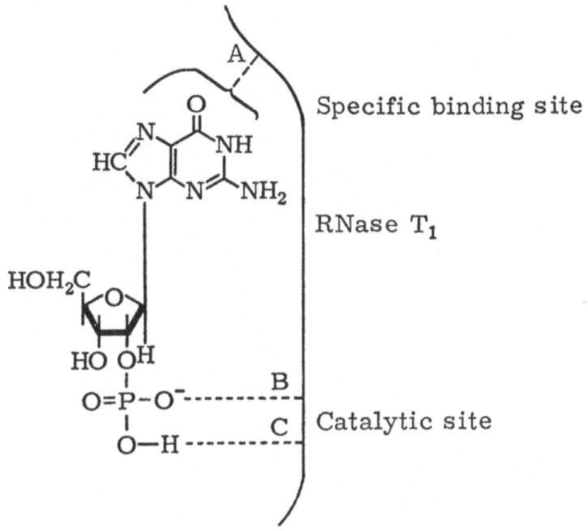

Chart IV-6. Binding of RNase $T_1$ and 2'-GMP

Summarising the observations on the specificity of the enzyme and on the interaction of the enzyme with 2'-guanylate, EGAMI suggested as a working hypothesis that the binding of RNase $T_1$ and 2'-guanylate might be as shown in Chart IV-6 (EGAMI, 1966 a, b).

## 2. RNase $T_2$

Ribonucleate nucleotido-2'-transferase (cyclizing), *Aspergillus oryzae*: tra, endo, $> p \rightarrow Xp$, Non-s, intra.

*Purification* (UCHIDA and EGAMI, 1966; UCHIDA, 1966). To a water extract of Takadiastase at pH 7.0 is added a small amount of DEAE-cellulose, which adsorbs most of the RNase $T_1$. The filtrate is heated at 80° for 2 min at pH 1.5 to 1.8. It is then subjected to ammonium sulfate fractionation. The fraction precipitated between 0.4 and 1.0 saturation with $(NH_4)_2SO_4$ is dissolved in pH 6.0 buffer, and dialyzed against water. Then the enzyme solution is subjected to DEAE-cellulose column chromatography. This process separates two fractions, RNase $T_2$-B and RNase $T_2$-A. Further purification is carried out by alcohol fractionation and repeated fractionation by DEAE-cellulose column chromatography. In this way RNases $T_2$-A and $T_2$-B are obtained in homogeneous states.

*Properties of RNase $T_2$* (UCHIDA and EGAMI, 1966; EGAMI et al., 1964). RNase $T_2$ digests RNA in two steps just like RNase $T_1$ (SATO and EGAMI, 1957; NAOI-TADA et al., 1959), but without absolute base specificity. It attacks all phosphodiester bonds in RNA with a preference for adenylic acid bonds (UCHIDA and EGAMI, 1967; RUSHIZKY and SOBER, 1963). The specificity of the enzyme will be described later.

The properties of highly purified RNase $T_2$ (UCHIDA, 1966) are as follows:

Table IV-9. *The properties of highly purified RNase $T_2$ are as follows*

| | |
|---|---|
| $S_{20,w}$ | $3.6_1$ S |
| molecular weight | |
| estimated from sedimentation equilibrium | 36,200 |
| Isoelectric point | ca. 5 |
| Absorption maximum | 281 mµ |
| Absorption minimum | 252 mµ |
| ODmax/ODmin | $2.5_2$ |
| $OD^{0.1\%}_{max\,1\,cm}$ | $1.9_9$ |
| Carbohydrate content | 12—15% |
| Amino terminal | Glutamic acid (or Glutamine) |
| Molecular activity | $15 \times 10^3$ |
| for uridine 2',3'-cyclic phosphate hydrolysis | |
| Optimum pH | |
| for RNA digestion | pH 4.5 |
| for uridine 2',3'-cyclic phosphate hydrolysis | pH 6.0—6.3 |
| Stability | |
| most stable around the neutrality: stable at pH 6.0, 80° for 5 min. | |
| more stable in alkaline medium and less stable in acidic medium than RNase $T_1$. | |

*Stability*: The enzyme is most stable around neutrality. It is stable at pH 6.0, 80° for 5 min. It is more stable in alkaline medium, and less stable in acidic medium than RNase $T_1$.

As enzymes and as proteins, there is little difference between the properties of RNase $T_2$-A and RNase $T_2$-B. The only difference between them, besides their behavior on DEAE-cellulose column chromatography, is in the nature of their sugar components. Thus RNase $T_1$-A contains mannose and glucose, while RNase $T_2$-B contains mannose and galactose as the main sugar components.

The effects of activators and inhibitors are summarized in Table IV-10 (UCHIDA, 1966).

In comparison with RNase $T_1$, RNase $T_2$ seems less sensitive to $Zn^{2+}$, $Ag^+$ and $Hg^{2+}$, and more sensitive to $Cu^{2+}$.

RNase $T_2$ may also be regarded as neither an SH-enzyme nor a serine enzyme. It requires no divalent cations.

*Amino acid composition of RNase T$_2$* (UCHIDA, 1966).

The characteristic features of the amino acid composition of RNase T$_2$ are as follows:

1. It contains six histidine residues per molecule, i.e. twice as many as RNase T$_1$. This is noteworthy, for some of these histidine residues may be involved in the catalytic function of ribonucleases. Photooxidation catalyzed by methylene blue also causes its inactivation.

2. It possesses both tryptophan (seven residues) and methionine (one residue).

Table IV-10. *Activators and inhibitors of RNase T$_2$*

| Reagents | Final concentration ($-\log M$) | Activity remaining (%) | | |
|---|---|---|---|---|
| | | RNase T$_2$-A | RNase T$_2$-B | RNase T$_1$[a] |
| NaCl | 0 | 40—50% | —% | 75% |
| | 1 | 90 | 90 | 100—115 |
| NaF | 1 | 104 | 107 | 93—100 |
| NaN$_3$ | 2 | 92 | 85 | 95—100 |
| Na$_2$S | 2 | 36 | — | 10 |
| AgNO$_3$ | 3 | 50 | — | 0 |
| MgCl$_2$ | 1 | 78 | 83 | 60 |
| CaCl$_2$ | 2 | 90 | 93 | 70—75 |
| HgCl$_2$ | 3 | 40 | — | 10 |
| MnSO$_4$ | 2 | 66 | — | 45 |
| ZnSO$_4$ | 3 | 50 | 50 | 0 |
| CuSO$_4$ | 3 | 0 | 0 | 20—50 |
| FeSO$_4$ | 2 | 58 | — | 20 |
| ICH$_2$COOH | 2 | 120 | 117 | 100 |
| BrCH$_2$COOH | 4 | 100 | — | 100 |
| DFP | 6 | 100 | — | 100 |
| Histidine | 2 | 100 | — | 82 |
| | 3 | 100 | — | 150 |
| EDTA | 2 | 110 | 110 | 125—150 |

Activity was measured at pH 4.5 with various inhibitors and activators instead of EDTA, according to assay for RNase T$_2$, but without EDTA. Addition of RNA, that is, the start of the reaction, was performed after preincubating enzyme and reagent for 20 min.

[a] Determined by TAKAHASHI.

A tryptophan residue is present in RNase T$_1$ but not in RNase I-A, while methionine residue is present in the latter but not in the former.

3. Eleven half cystine residues but no free-SH-groups have been found.

4. Basic amino acids constitute about 10% of the total amino acid residues. This high content of basic amino acids, and especially of lysine, distinguishes RNase T$_2$ from RNase T$_1$ which is very poor in basic amino acids other than histidine.

5. Like RNase T$_1$ it has a higher content of glycine than of alanine.

6. It has relatively many alanine residues. Generally speaking, the amino acid composition of RNase T$_2$ is quite different from that of RNase T$_1$.

Analysis of the N-terminal amino acid revealed that one molecule of RNase T$_2$ contains one residue of glutamic acid or glutamine. So it is highly probable that it consists of a single polypeptide chain like RNase I-A and RNase T$_1$.

*Specificity of RNase $T_2$.* In 1957 it was found that partially purified RNase $T_2$ preferentially attacks the adenylic phosphodiester bonds in RNA (SATO and EGAMI, 1957; NAOI-TADA et al., 1959). So by analogy to RNase $T_1$, which is specific for guanylic acid phosphodiester bonds, completely pure RNase $T_2$ was expected to be specific for adenylic acid phosphodiester bonds. However, contrary to expectation, homogeneous preparations of RNase $T_2$ were found to attack practically all phos-

Table IV-11. *Amino acid composition of RNase $T_2$*

| Amino acid | Found (Molar ratio as Arg = 1) | | | | Residues per molecule | |
|---|---|---|---|---|---|---|
| | 32 h hydrolysate | 72 h hydrolysate | Oxidized RNase $T_2$ | Corrected value | Molar ratio as Met = 1 | Expressed in nearest integer |
| Aspartic acid | 10.0 | 10.0 | $7.9_5$ | 10.0 | $38._9$ | 39 |
| Threonine | $5.7_0$ | $4.6_6$ | $4.1_6$ | $6.5_0$[b] | $25._3$ | 25 |
| Serine | $7.1_5$ | $5.9_5$ | $5.9_2$ | $8.1_0$[b] | $31._5$ | 32 |
| Glutamic acid | $10._1$ | $10._1$ | $8.1_9$ | $10._1$ | $39._3$ | 39 |
| Proline | $5.5_6$ | $5.9_5$ | $4.7_0$ | $5.9_5$[c] | $23._2$ | 23 |
| Glycine | $7.0_9$ | $7.2_9$ | $6.0_0$ | $7.1_9$ | $28._0$ | 28 |
| Alanine | $4.6_3$ | $4.8_6$ | $3.8_5$ | $4.7_5$ | $18._5$ | 19 |
| Cysteic acid | $2.5_8$ | $1.7_1$ | $2.5_6; 2.9_7$ | $2.7_5$[a] | $10._6$ | 11 |
| Valine | $1.5_2$ | $1.6_7$ | $1.2_3$ | $1.6_7$[c] | $6._5$ | 7 |
| Methionine Sulfone | — | — | $0.2_4; 0.2_7$ | $0.2_6$[a] | $1._0$ | 1 |
| Isoleucine | $4.6_3$ | $4.9_1$ | $3.7_4$ | $4.9_1$[c] | $19._1$ | 19 |
| Leucine | $4.6_3$ | $4.7_1$ | $3.7_4$ | $4.7_1$[c] | $18._3$ | 18 |
| Tyrosine | $3.6_7$ | $3.6_2$ | $3.0_3$ | $3.7_0$[b] | $14._4, 14._7$[e] | 14 |
| Rhenylalanine | $2.5_8$ | $2.5_2$ | $2.2_3; 2.5_5$ | $2.5_5$ | $9._9$ | 10 |
| Lysine | $6.0_2$ | $6.0_1$ | $4.6_4$ | $6.0_2$ | $23._4$ | 23 |
| Histidine | $1.3_9$ | $1.5_2$ | $1.0_5$ | $1.4_6$ | $5._7$ | 6 |
| Arginine | 1.0 | 1.0 | 1.0 | 1.0 | $3._9$ | 4 |
| Tryptophan | — | — | — | — | $—, 7._3$[e] | 7 |
| (Amide $NH_3$) | $18._8$ | $21._9$ | $20._8$ | $16._3$[d] | $63._2$ | (63) |
| Total | | | | | | 325 |

[a] The amounts of cysteic acid and methionine sulfone were expressed as mean values between the molar ratio calculated as Arg = 1.0 and another ratio as Phe = $2.5_5$.

[b] The results for the labile amino acids (Ser, Thr) were corrected by extrapolating to time 0.

[c] Value for the 72 h hydrolysate only.

[d] Extrapolated to time 0.

[e] Determined spectrophotometrically.

phodiester bonds in RNA with a preference for adenylic acid bonds (UCHIDA and EGAMI, 1967). Similar observations were reported by RUSHIZKY and SOBER (1963).

The specificity of RNase $T_2$ was studied by UCHIDA and EGAMI using mainly RNA's and polyribonucleotides as substrates (UCHIDA and EGAMI, 1967), by RUSHIZKY and SOBER using mainly oligoribonucleotides (RUSHIZKY and SOBER, 1963), by SATO et al. with nucleoside 2',3'-cyclic phosphates (SATO et al., 1966) and by M. TADA with special reference to the susceptibility of methylated nucleotide phosphodiester bonds in RNA (TADA, 1966).

The results may be summarized as follows (UCHIDA and EGAMI, 1967):

1. Although RNase T$_2$ shows a preference for the adenylic acid linkage in RNA, it has no absolute specificity as regards nucleotide bases and cleaves practically all nucleotide bonds in RNA and polyribonucleotides.

2. In exhaustive digestion with RNase T$_2$, RNA is completely digested to produce nucleoside 3'-phosphates.

3. When estimated by the production of acid soluble nucleotides, the time course of digestion of high-molecular weight RNA by RNase T$_2$ apparently proceeds in three phases: "an initial slow phase", "a rapid phase" and "a second slow phase". This is because the digestion was estimated, not by the rate of cleavage of internucleotide bonds, but by the increase in the optical density of the acid soluble products. So it may be inferred that in "the initial slow phase" RNA was split into relatively large oligonucleotides, most of which were acid-insoluble and had adenylate at the 3'-terminal.

4. The susceptibility to RNase T$_2$ depends on the nature of RNA and various homopolynucleotides. Poly A is fairly resistant to RNase T$_2$ in spite of the preferential cleavage of adenylic acid linkages in RNA, whereas unexpectedly poly U is the most sensitive to the enzyme of the various polynucleotides tested. The relative resistance of poly A to the enzyme may be explained by the formation of a rigid double-stranded helical structure in acidic medium (UCHIDA and EGAMI, 1967). Poly C is also poorly digested by RNase T$_2$. Poly U, which does not form this type of rigid conformation, is easily digested by the enzyme. s-RNA, containing much double-stranded structure, is more resistant to RNase T$_2$ than commercial yeast RNA. The latter is more sensitive to the enzyme than high molecular yeast RNA.

5. As intermediates of RNA digestion, $G > p$, $U > p$ and $C > p$ are accumulated, but $A > p$ scarcely accumulates. As intermediates of poly U digestion, UpUp and $U > p$ are detected. This means that RNase T$_2$ behaves as an endonuclease for poly U as for RNA.

In 40% methanol solution, where part of its rigid double-stranded structure is destroyed, poly A is more susceptible to RNase T$_2$ than in water and as intermediates of digestion, ApAp and ApApAp, are detected (UCHIDA and EGAMI, 1967).

6. Results essentially consistent with those mentioned above were obtained by RUSHIZKY and SOBER using RNA and various trinucleotides as substrates (RUSHIZKY and SOBER, 1963).

7. Although RNase T$_2$ had no absolute base specificity, it was found that $\beta$-methyl riboside 2',3'-cyclic phosphate was not cleaved by RNase T$_2$ (UCHIDA and EGAMI, 1967). So at least part of the base structure seems to be of significance for the reaction by RNase T$_2$.

Thus the Ap·N-oxide phosphodiester bonds in RNA N-oxide, prepared by the action of perphthalate on RNA were found to be resistant to RNase T$_2$.

8. Concerning the minor components in s-RNA, MADISON and HOLLEY have shown that RNase T$_2$ is able to split the phosphodiester bonds of pseudouridylic acid and 5,6-dihydrouridylic acid (MADISON and HOLLEY, 1965).

According to M. TADA, unlike the phosphodiester bonds of 5-methyl Gp, $\Psi$p, and dihydro Up, the phosphodiester bonds of $N_{(2)}$-dimethyl Gp and 1-methyl Gp are fairly resistant to RNase T$_2$ (1966). Larger amounts of RNase T$_2$ may be required to cleave these phosphodiester bonds completely.

9. SATO et al. (1966) studied the action of RNase $T_2$ on various nucleoside 2′,3′-cyclic phosphates and determined the $Km$ and $V_{max}$ at the optimum pH (pH 6.0).

Table IV-12. *Michaelis constant and maximum velocity for RNase $T_2$ hydrolysis of cyclic nucleotides measured by the titration method*

| Cyclic nucleotide | $Km$ (M) | $Vm$ (moles/min/ mg protein) |
|---|---|---|
| A-cyclic-p | $2.3 \times 10^{-4}$ | $3.6 \times 10^{-4}$ |
| C-cyclic-p | $4.0 \times 10^{-4}$ | $3.5 \times 10^{-4}$ |
| G-cyclic-p | $6.5 \times 10^{-4}$ | $2.5 \times 10^{-4}$ |
| U-cyclic-p | $6.1 \times 10^{-4}$ | $4.4 \times 10^{-4}$ |
|  | $4.4 \times 10^{-4}$[a] | $4.2 \times 10^{-4}$[a] |

[a] By the spectrophotometric method.

They also measured the $Ki$ values of nucleotides which competitively inhibited the reaction. The results are summarized in Table IV-13.

Table IV-13. *Inhibitor constants between RNase $T_2$ and nucleotides measured in hydrolysis of uridine cyclic phosphate by the spectrophotometric method*

| Inhibitor | $Ki$ (M) |
|---|---|
| 2′-AMP | $1.1 \times 10^{-5}$ |
| 2′-CMP | $1.2 \times 10^{-4}$ |
| 2′-GMP | $3.4 \times 10^{-4}$ |
| 3′-AMP | $3.8 \times 10^{-5}$ |

RNase $T_2$ attacks RNA with a preference for adenylic acid bonds. However, as far as the $Km$ value and $V_{max}$ for cyclic nucleotide hydrolysis are concerned, no remarkable difference was observed between the values with A-cyclic-p and those with other nucleotides. The $V_{max}$ value for A-cyclic-p hydrolysis is even smaller than that for U-cyclic-p hydrolysis. So the apparent preference for adenylic-acid bonds observed in RNA digestion might be due to the base specific affinity of RNase $T_2$ to adenylate. The finding that the $Ki$ values for 2′-AMP are much smaller than those for other nucleotides supports this consideration.

## C. RNases of Streptomyces and Actinomyces

A ribonuclease with a similar specificity to that of RNase $T_1$ was found by TANAKA in the culture medium of *Streptomyces erythreus* (TANAKA, 1961). A similar enzyme has also been found in *Streptomyces albogriceolus* by YONEDA (1964) and in *Actinomyces aureoverticillatus* by N. M. ABRUSIMOVA-AMELYANCHIK et al. (1965).

## 1. RNase of *Streptomyces erythreus*

[Ribonucleate guaninenucleotido-2'-transferase (cyclizing)]; tra, endo, $> p \rightarrow Xp$,
G-s. extra (TANAKA, 1961)

*Streptomyces erythreus* (American Type Culture Collection, No. 11635 or Northern Research Laboratory No. 2338) cells grown in medium, with glucose, Difco-Bacto-yeast extract, Difco-Bacto-beef extract, and Bacto-casamino acids at 32 °C for 48 h are inoculated into water medium containing soy bean meal, potato starch, glycerol, corn-steep liquor, yeast extract, NaCl and $CaCO_3$ at pH 7.0 and incubated for 5 days. Then the mycelia are centrifuged off and RNase is purified from the supernatant. RNase is precipitated from the supernatant with acrinol (Rivanol, 2-ethoxy-6, 9-diamino-acridinium lactate). This means that RNase behaves as an acidic protein. The yellow precipitate is extracted with 1 *M* sodium acetate. RNase is precipitated from the extract with cold acetone. The precipitate of crude RNase is washed with acetone and dried under reduced pressure. The acetone powder can be stored in a desiccator at low temperature for several years without significant loss of activity.

For further purification, the acetone powder is dissolved in water and RNase is adsorbed on DEAE-cellulose equilibrated with 0.1 *M* Na-acetate buffer, pH 4.0, by batch-wise treatment. RNase is eluted with 0.06 *M* NaCl in 0.1 *M* acetone buffer, pH 4.0. The active effluents are concentrated under reduced pressure at low temperature and dialyzed against distilled water. The enzyme is further purified by repeated DEAE-cellulose column chromatography. The active fractions are pooled and stored in a deep freezer or lyophilized and stored in a desiccator. The purified preparation has a specific activity (units/$A_{280\,m\mu}$) of about 1,000 times more than the starting broth.

*pH Optimum*: The optimum pH of this enzyme is 7.3 to 7.4 in 0.2 *M* phosphate buffer.

*Heat-stability*: The purified enzyme shows remarkable stability. It is not inactivated at all by treatment at 80° and pH 5 for 5 min. Even on treatment at 100° for 10 min., it retains 85, 90 and 51% of its activity at pH 7.4, 5.6 and 2.0, respectively.

*Specificity* (TANAKA, 1961; TANAKA, 1966; TANAKA and CANTONI, 1963): Digestion of yeast RNA with this enzyme preparation yields 3'-GMP and oligonucleotides terminated with 3'-GMP via terminal 2',3'-cyclic phosphates. Among 2',3'-cyclic nucleotides, only guanosine 2',3'-phosphate is hydrolyzed yielding 3'-GMP. These findings indicate that the specificity of this enzyme is quite similar to that of RNase $T_1$. However it remains to be seen, if there is indeed a slight difference in specificity between this enzyme and RNase $T_1$ with regard to minor components in s-RNA, as suggested by TANAKA and CANTONI (1963).

It should be added here that on DEAE-cellulose column chromatography two peaks were found: a major peak (A) and a minor peak (B). So far (Optimum pH, specificity), no difference has been found between them. TANAKA suggested that the minor component (B) might be derived from the enzyme in the major component (A) by the action of a proteolytic enzyme in the culture broth. However other possibilities could not be excluded as in the case of pancreatic RNases (RNase I-A and RNase I-B) and Ribonucleases $T_2$-A and $T_2$-B.

## 2. RNase of Streptomyces albogriceolus

[Ribonucleate guaninenucleotido-2′-transferase (cyclizing)]; tra, endo, $> p \rightarrow Xp$, G-s, extra (YONEDA, 1964).

M. YONEDA has partially purified an RNase $T_1$-like enzyme from the culture medium of *Streptomyces albogriceolus* (YONEDA, 1964).

*Properties*: The enzyme is resistant to heat-treatment at 80 °C, for 5 min at various pH values. It is quite resistant at pH 3—4, but the activity decrea- ses at above pH 5. It is fairly unstable in alkaline medium (pH 8) even at 37 °C.

The effects of various metal ions and other effectors on it are shown in Table IV-14.

Table IV-14. *Ribonuclease from streptomyces albogriceolus*
Stability of streptomyces ribonuclease remaining activity (%)

| pH | 80 °C | 37 °C | | | |
|---|---|---|---|---|---|
| | 5 min | 30 min | 3 h | 4.5 h | 22 h |
| 3 | 100 | | | | |
| 4 | 100 | 100 | 100 | 100 | 100 |
| 5 | 28 | 98 | 98 | 97 | 76 |
| 6 | 20 | 97 | 95 | 95 | 35 |
| 7 | 12 | 95 | 92 | 92 | 15 |
| 8 | 21 | 93 | 75 | 70 | 3 |
| 9 | 32 | | | | |

Stability was tested on the mixture containing 1 ml of enzymes and equal volume of each buffer. At appropriate time, mixture was tested for ribonuclease activity. Remaining activity as compared with the initial activity was shown as %.

Effect of various cations and anions on ribonuclease fraction

| Additions | Ratio of activity $10^{-2} M$ | $10^{-3} M$ | $10^{-4} M$ |
|---|---|---|---|
| CuSO$_4$ | 0.18 | 0.93 | 0.96 |
| MgSO$_4$ | 0.76 | 0.90 | 1.02 |
| CaCl$_2$ | 0.79 | 0.99 | 1.03 |
| FeSO$_4$ | 0.37 | 0.55 | 1.02 |
| (NH$_4$)$_2$SO$_4$ | 0.96 | 1.01 | 1.04 |
| NiSO$_4$ | 0.28 | 0.80 | 0.94 |
| CoCl$_2$ | 0.43 | 0.95 | 0.81 |
| ZnSO$_4$ | 0.23 | 0.67 | 0.97 |
| MnCl$_2$ | 0.56 | 0.78 | 0.96 |
| K$_2$HPO$_4$ | 0.99 | 1.00 | 0.94 |
| EDTA | 1.18 | 1.15 | 1.17 |
| Na-Acetate | 1.02 | 1.09 | 0.96 |
| Na-Succinate | 1.08 | 1.03 | 1.08 |
| Na-Citrate | 1.12 | 1.09 | 1.06 |

See text about activity measurement. The numerals express the ratio of ribonuclease activity in the presence or absence of additional compounds.

It does not require divalent cations. Metal ions such as $Cu^{2+}$, $Fe^{2+}$, and $Zn^{2+}$ are slightly inhibitory. EDTA has no effect.

*Specificity*: From 3′-terminal analysis of the RNA digestion products formed by the RNase, it is concluded that it cleaves RNA at the guanylate linkage to produce 2′,3′-cyclic guanylic acid which is further hydrolyzed to 3′-guanylic acid, just like RNase $T_1$.

This finding together with others shows that guanyloribonucleases are fairly widely distributed in microorganisms, both fungi and bacteria.

### 3. RNase of *Actinomyces aureoverticillatus*

[Ribonucleate guanine-nucleotido-2′-transferase (cyclizing)]; tra, endo, $> p \rightarrow pX$, G-s, extra.

N. M. ABROSIMOVA-AMELYANCHIK, R. I. TATARSKAYA et al. isolated a guanyloribonuclease from the culture fluid of *Actinomyces aureoverticillatus* KRAS. et DISHEN, strain 1306 (1965). They found the cleavage of RNA by this enzyme proceeds in two steps just like that by RNase $T_1$. To clarify the specificity of the enzyme, oligonucleotides with minor components in s-RNA were prepared from s-RNA digestion products and subjected to the action of the enzyme. It was shown that the following cleavages occur.

$$GpUp \rightarrow G > p + Up \rightarrow Gp + Up$$

$$(\text{1-Methyl Gp, Gp}) \, Cp \rightarrow \text{1-Methyl G} > p + G > p + Cp$$

$$\begin{array}{c} \downarrow \quad\quad\quad\quad \downarrow \\ \text{very} \quad Gp \\ \text{slowly} \\ \text{1-Methyl Gp} \end{array}$$

$$(\text{1-Methyl Gp, } N_{(2)}\text{-Methyl Gp}) \, Cp \rightarrow \text{1-Methyl G} > p + N_{(2)}\text{-Methyl G} > p + Cp$$

$$\begin{array}{c} \downarrow \quad\quad\quad\quad \downarrow \\ \text{very} \quad N_{(2)}\text{-Methyl Gp} \\ \text{slowly} \\ \text{1-Methyl Gp} \end{array}$$

$$(\text{1-Methyl Gp, } N_{(2)}\text{-Dimethyl Gp}) \, Cp \rightarrow \text{1-Methyl G} > p + N_{(2)}\text{-dimethyl G} > p + Cp$$

$$\begin{array}{c} \downarrow \quad\quad \text{very} \quad\quad\quad \downarrow \quad\quad \text{very} \\ \text{slowly} \quad\quad\quad\quad \text{slowly} \\ \text{1-Methyl Gp} \quad\quad N_{(2)}\text{-dimethyl Gp} \end{array}$$

The enzyme isolated from *Actinomyces aureoverticillatus* cleaves all the bonds formed by methylated derivatives of guanine frequently encountered in s-RNA.

They further compared the rates of cleavage of various dinucleotides by the enzyme. The rate of cleavage decreases if the first nucleotide, guanylate, is replaced by $N_{(2)}$-dimethyl guanylate and xanthylate. The rate of cleavage of dinucleotides also depends on the nature of the second nucleotide; e.g.

$$Gp \, \Psi p < GpUp \, .$$

From these results it is concluded that for the reaction a purine base with a hydroxy group at $C_{(6)}$ must be present in the nucleotide chain of RNA. However, taking into consideration the state of the guanine residue at the optimum pH (7.6) of the enzyme and the susceptibility of 1-methyl Gp phosphodiester bonds to the enzyme, we consider that the actual structure of the substrate required is an oxo group at $C_{(6)}$ of the purine base, as in the case of RNase $T_1$.

The important rôle of a proton in $C_{(1)}$ and the effects of substitution at $C_{(2)}$ (amino in G, oxo in X, and hydrogen in I) seem to be at least quantitatively, the same as for RNase $T_1$. However, as ABROSINOVA-AMELYANCHIK et al. suggest, from a comparison of their results with these of STAEHELIN, the proton in the $C_{(1)}$-position plays a greater rôle in RNase $T_1$ than in this enzyme.

Moreover, it should be added here that there is a difference in the effects of nearest neighbor residues on the cleavage rate by this enzyme and by RNase $T_1$. Thus, GpC is cleaved approximately half as rapidly as GpU, while, according to WHITFELD and WITZEL, GpC is cleaved approximately three times as rapidly as GpU by RNase $T_1$. These findings are significant not only for elucidation of the relationship between the structure and function of guanyloribonucleases, but also for the application of these enzymes to nucleotide sequence analysis of RNA. In this connection the recent finding by KOCHETKOW et al. is noteworthy (1967). They reported that chemical modification of s-RNA by glyoxal followed by the action of RNase $T_1$ allows splitting of the polynucleotide chain at inosine and 2-dimethylamino-6-oxopurine units, and gives rise to large fragments, while the action of this enzyme on modified s-RNA gives rise to small fragments owing to the susceptibility of glyoxal-modified guanylate bonds, unlike that of RNase $T_1$.

# D. RNases of Ustilago

## 1. RNases of Ustilago sphaerogena and Ustilago zeae

GLITZ and DEKKER found that when grown on a medium in which the sole source of carbon is RNA, the smut fungus *Ustilago sphaerogena* produces an extracellular guanyloribonuclease (GLITZ and DEKKER, 1964).

Similar phenomena have been observed by YANAGIDA, UCHIDA and EGAMI (1964) who have found that when grown on a medium in which the sole source of phosphorus is RNA, *Ustilago zeae* produces several times as much extracellular guanyloribonuclease as in normal medium. They further observed the curious phenomenon that even poly U enhances the production of the guanyloribonuclease (YANAGIDA et al., 1964).

These interesting findings with *Ustilago* led ARIMA, UCHIDA and EGAMI to investigate the mode of formation and the nature of RNases of *Ustilago sphaerogena* further. Besides the guanyloribonuclease (RNase $U_1$) described by GLITZ and DEKKER they found a novel enzyme, purine specific RNases (RNases $U_2$ and $U_3$) and a nonspecific RNase (RNase $U_4$) (ARIMA, UCHIDA and EGAMI, 1968). The properties of these enzymes are as follows:

a) Guanyloribonuclease of *Ustilago sphaerogena* investigated by GLITZ and DEKKER. *Ustilago sphaerogena* was grown on a medium in which the sole source of carbon was RNA, and ribonuclease (Guanyloribonuclease) was purified from the culture medium about 300 fold. It had a specific activity equivalent to that of other highly purified ribonucleases and was essentially homogeneous upon ultracentrifugation and electrophoresis.

*Properties*: Optimum pH, varies with the buffer. The highest activity is found with imidazole buffer (sharp optimum at pH 7.0—7.5). Activity is higher at pH 7.0 with imidazole and Tris buffers than at pH 6.0, and the reverse is true with bicarbonate buffer. The enzyme requires a protective protein, such as bovine plasma albumin. In its absence the observed activities are low and hard to reproduce. No organic or metallic cofactor is necessary. In 0.1 $M$ NaCl or $5 \times 10^{-4}$ $M$ MgCl$_2$ the rate of RNA digestion by the *Ustilago* RNase is about 50% of the maximal value. The effects are less remarkable for G > p hydrolysis. Unlike pancreatic RNase (RNase I-A), the enzyme is not inhibited by polyvinyl sulfate.

This may be explained by the acidic nature of the enzyme similar to that of RNase T$_1$. Spermine and spermidine strongly inhibit RNA digestion by the enzyme. The enzyme appears to be very stable to a number of possible denaturants. Heating at a temperature of 100° for at least 5 min results in no loss of enzyme activity. Neither iodoacetate at pH 6 nor *p*-CMB inhibit the enzyme, showing that it is not an SH-enzyme.

The sedimentation constant is ca. 1.6 S. Most of the properties of the *Ustilago* enzyme are quite similar to those of RNase T$_1$. The differences to be noted are: 1. The requirement of the former for a protective protein, and 2. The strong inhibition of the former by chloride ion.

*Specificity* (GLITZ and DEKKER, 1964): The extracellular RNase of *Ustilago sphaerogena* may be regarded with some reserve as a ribonucleate guanine-nucleotido-2′-transferase (cyclizing) 2.7.7.26. Hydrolysis of RNA by the RNase results in the formation of 3′-guanylic acid and oligonucleotide ending in guanosine 3′-phosphate. The enzyme first acts through the formation of guanosine 2′,3′-cyclic phosphate, and then hydrolyzes the cyclic ester more slowly to form the 3′-nucleotide. The hydrolysis of terminal guanosine 2′,3′-phosphate of oligonucleotides is more rapid than that of guanosine 2′,3′-cyclic phosphate to form 3′-guanylic acid. A copolymer of adenylic acid and guanylic acid is attacked at all guanylic acid residues, resulting in the formation of G > p, Gp and a series of oligonucleotides composed of adenylic acid with 3′-terminal guanylic acid. Polyadenylic acid is scarcely hydrolyzed by the enzyme, while the cleavage of poly AI by the enzyme is somewhat surprising. Contrary to the results expected from the specificity of the enzyme, cleavage occurs adjacent to both inosinic acid and adenylic acid residues. If the substrate specificity of the enzyme were essentially the same as that of RNase T$_1$, only the inosinic acid phosphodiester bonds in poly AI should be cleaved. Although GLITZ and DEKKER suggested two possibilities for the mechanism, no convincing explanation was given. The possibility of slight contamination with "purine-specific RNase" (see below) cannot be excluded.

Poly U, Poly C, DNA and bis-(*p*-nitrophenyl) phosphates are quite inert to the enzyme.

b) RNases U$_1$, U$_2$, U$_3$ and U$_4$ described by ARIMA, UCHIDA and EGAMI (1968).

When *Ustilago sphaerogena* is cultured in a medium containing glucose, glycine and mineral salts, four ribonucleases are released into the culture medium. They are characterized as follows:

| Individual name | Systematic name | EC number | Trivial name | Characters |
|---|---|---|---|---|
| RNase $U_1$ | Ribonucleate guaninenucleotido-2'-transferase (cyclizing) | 2.7.7.26 | Guanyloribonuclease | tra, endo, $> p \rightarrow Xp, G\text{-s}$, extra. |
| RNase $U_2$ | Ribonucleate purinenucleotido-2'-transferase (cyclizing) | | Puryloribonuclease | tra, endo, $> p \rightarrow Xp$, Pur-s, extra. |
| RNase $U_3$ | Ribonucleate purinenucleotido-2'-transferase (cyclizing) | | Puryloribonuclease | tra, endo, $> p \rightarrow Xp$, Pur-s, extra. |
| RNase $U_4$ | | | | Xp Non-s, extra. |

The production of RNases $U_1$ and $U_4$ is enhanced by the addition of RNA to the culture medium[1] (ARIMA, UCHIDA and EGAMI, 1968).

*Purification*: DEAE-cellulose is added to the culture filtrate at pH 8.5 and then filtered off. RNases are eluted from the filter cake with 0.3 M NaCl solution at pH 8.5. The enzyme solution, after dialysis, is adsorbed on CM-cellulose at pH 4.0. Then the enzymes are eluted from the CM-cellulose with 0.2 M NaCl solution at pH 4.0. The filtrate is dialyzed and subjected to DEAE-cellulose column chromatography at pH 8.5 with an increasing concentration of NaCl and three peaks with RNase activity corresponding to RNases $U_1$, $U_2$, and $U_3$, are eluted successively.

For the purification of RNase $U_4$, *Ustilago sphaerogena* is cultured in RNA-medium, where RNA is added as a sole source of phosphorus instead of phosphate. The enzyme is purified by ammonium sulfate precipitation, Sephadex G-75 gelfiltration and DEAE-cellulose column chromatography.

In this way RNases $U_1$, $U_2$, $U_3$ and $U_4$ are purified about 1600, 3700, 1100 and 16-fold, respectively, from the culture filtrate.

*Properties of RNases $U_1$, $U_2$, $U_3$ and $U_4$*: It was shown by Sephadex G-75 gel filtration that RNases $U_1$, $U_2$ and $U_3$ have molecular weights of about 10,000 while RNase $U_4$ is much larger. The optimum pH values of RNases $U_1$ and $U_4$ are about pH 8 and those of RNases $U_2$ and $U_3$ are about pH 4.5. Unlike the RNase described by GLITZ and DEKKER, none of these enzymes needs protective protein for activity even at the final stage of purification.

The effects of inhibitors and activators on these enzymes are summarized in Table IV-15.

No metallic cofactor is necessary for the actions of these RNases. $AgNO_3$ and $CuSO_4$ are inhibitory to all four enzymes. $Zn^{2+}$ is inhibitory to RNases $U_1$ and $U_4$. $Mg^{2+}$ and $Mn^{2+}$ inhibit RNase $U_4$. All these inhibitions, except that by $AgNO_3$, are reversed by EDTA. Since neither iodoacetate nor *p*-chloromercuribenzoate has any effect on these RNases, it appears that they are not SH-enzymes.

RNases $U_1$, $U_2$ and $U_3$ are as stable as RNase $T_1$. These three RNases are fairly thermostable: no loss of activity occurs on heating at 80° and pH 6.9 for 4 min, but RNase $U_4$ is as heat labile as most other enzymes.

---

[1] Details will be published elsewhere.

*Substrate specificity*: RNase $U_1$ splits only the phosphodiester bonds of guanosine 3'-phosphates in RNA. It may be regarded as a guanyloribonuclease [EC 2.7.7.26, ribonucleate guaninenucleotido-2'-transferase (cyclizing)] similar to RNase $T_1$. It seems to be identical with the extracellular RNase described by GLITZ and DEKKER (1964).

RNases $U_2$ and $U_3$ are novel enzymes with strict specificity. They split the inter-nucleotide bonds between 3'-purine nucleotides in RNA with the intermediary formation of purine nucleoside 2',3'-phosphates, which are slowly hydrolyzed to

Table IV-15. *Inhibitors and activators of RNases from Ustilago sphaerogena*

| Reagent | Final concentration ($-\log M$) | Remaining activity (%) | | | |
|---|---|---|---|---|---|
| | | RNase $U_1$ | RNase $U_2$ | RNase $U_3$ | RNase $U_4$ |
| NaCl | 2 | 83 | 71 | 85 | — |
| | 3 | 107 | 104 | 106 | 100 |
| AgNO$_3$ | 2 | 5 | 14 | 28 | — |
| | 3 | 60 | 60 | 61 | 12 |
| MgSO$_4$ | 2 | 33 | 79 | 70 | — |
| | 3 | 96 | 110 | 113 | 21 |
| CaCl$_2$ | 2 | 56 | 94 | 113 | — |
| | 3 | 98 | 119 | 132 | 62 |
| MnCl$_2$ | 2 | 24 | 72 | 83 | — |
| | 3 | 104 | 158 | 166 | 54 |
| CuSO$_4$ | 2 | 6 | 5 | 10 | — |
| | 3 | 61 | 53 | 72 | 6 |
| FeSO$_4$ | 3 | 109 | 112 | 112 | 102 |
| ICH$_2$COOH | 2 | 117 | 55 | 47 | — |
| | 3 | 108 | 103 | 93 | 65 |
| EDTA | 2 | 115 | 91 | 112 | — |
| | 3 | 113 | 100 | 102 | 479 |
| Zn(CH$_3$CO$_2$)$_2$ | 2 | 4 | 56 | 72 | — |
| | 3 | 41 | 102 | 111 | 32 |
| PCMB | 4 | 101 | 95 | 115 | 100 |
| None | | 100 | 100 | 100 | 100 |

RNase activity was measured at the optimal pH by the usual assay procedure.

3'-purine nucleotides. So they may be classified as "puryloribonuclease [ribonucleate purinenucleotido-2'-transferase (cyclizing)]". Double stranded RNA is resistant to RNases $U_2$ and $U_3$.

It should be noted here that the preparations of RNases $U_2$ and $U_3$ may be contaminated by a small amount of other enzymes which attack RNA. Both preparations attack adenylate and guanylate bonds in RNA with a preference for the former. So the possibility, that one or both of them is adenine specific RNase, has not been excluded.

RNase $U_4$ has no base specificity. The products of exhaustive digestion of RNA by the enzyme are 3'-adenylate, 3'-guanylate, 3'-cytidylate and 3'-uridylate, and no small oligonucleotides can be detected during the digestion. This suggests that a non-specific exonuclease may produce 3′-nucleotides (hyd, exo, → Xp, Non-s). Further studies are required to elucidate this possibility.

## E. RNases of Neurospora crassa

Although the metabolic pathways and enzymes of *Neurospora crassa* have been extensively investigated from the point of view of biochemical genetics, there are few reports on the ribonucleases of this organism. SUSKIND and BONNER extracted an intracellular RNase and described its enzymatic properties (1960). TAKAI, UCHIDA and EGAMI described an extracellular RNase (RNase $N_1$) and two intracellular RNases (RNase $N_2$ and RNase $N_3$) (1966, 1967).

a) Intracellular RNase found by SUSKIND and BONNER

SUSKIND and BONNER carried out studies to see whether gene mutation in *Neurospora crassa* elicited genetic changes in nucleic acid or protein formation.

Electrophoretic and biochemical analysis of a water extract of lyophilized mycelia of wild and mutant strains did not show any specific alteration attributable to gene mutation. In the course of this study they characterized an intracellular RNase.

*Extraction*: All wild type and morphological mutant cultures and biochemical mutant cultures were grown on a minimal medium plus glucose or sucrose and on minimal medium supplemented with the appropriate nutrient, respectively. After growth, the mycelia were harvested and lyophilized. The lyophilized mycelia were ground to a fine powder in a mortar. The powder was extracted with phosphate buffer, at pH 7.5 to 7.8 by shaking with glass. Ultracentrifugation of the crude extract at $41,190 \times g$ for 90 min did not precipitate ribonuclease activity.

*Properties*: Purine and pyrimidine bases, nucleosides, and nucleotides do not inhibit the enzyme at concentrations as high as $10^{-2}\ M$, with the exception of ATP which is inhibitory at $8 \times 10^{-3}\ M$.

Maximum activity is found at pH 7.5 with about 75% loss of activity at pH 6.0. Total inactivation occurs at pH 9.2. An increase in phosphate from 0.2 $M$ to 0.5 $M$ results in a loss of activity. Maximum activity is found at 60° in 0.1 $M$ phosphate buffer. The enzyme is resistant to elevated temperature for a prolonged period of time.

Dialysis of the enzyme preparation causes no loss in activity. Almost all of the divalent ions tested at concentrations of $10^{-2}\ M$ caused inhibition of the enzyme. The most effective inhibitors at $10^{-3}\ M$ were copper and cobalt. Both ferrous and ferric ions are inhibitory. The possibility of inhibiting the enzyme with EDTA at several pH values was examined. The inhibitory effect of EDTA is greatest at pH 7.0. At 0.1 $M$ EDTA, no significant RNase activity remains. The inhibitory effect of EDTA can be almost completely reversed by dialysis against 0.05 $M$ phosphate buffer, pH 7.8, for 24 h.

SUSKIND and BONNER estimated intracellular RNase activity per pad and total extracellular RNase activity in the filtered medium during growth of the wild, and various mutant strains. Activity was found in the filtrate, and generally the ratio of extracellular/intracellular activity increased with the age of the culture.

b) RNase $N_1$, $N_2$ and $N_3$

The formation of ribonucleases (RNase), phosphodiesterases (PDase), and phosphomonoesterases (PMase) was studied by TAKAI, UCHIDA and EGAMI (1966, 1967) in cultures of three different strains of *Neurospora crassa* (wild strain 74 A, and two adenine-requiring strains, 74A-T28-M4 and 74A-Y152-M7ylo) in various culture media (Medium A: Ordinary medium; Medium B: RNA without terminal P and KCl in place of $KH_2PO_4$ of medium A as the source of P and K; Medium C: Medium A supplemented with adenine; Medium D: Medium A supplemented with RNA. Mediums C and D were used for adenine-requiring mutants). The following results were found:

1. *Neurospora crassa*, when cultured in Medium A reached the logarithmic phase of growth after about 10 h and stopped growing after about 50 h. The culture released an extracellular RNase (RNase $N_1$, EC 2.7.7.26, guanyloribonuclease) into the medium only during the stationary phase.

2. No difference in extracellular RNase activity was observed between the wild strain culture in Medium A and in Medium B. The same was also found with the adenine-requiring strains cultured in Medium C and Medium D. So, unlike the case with *Ustilago*, it was concluded that RNA did not induce the formation or release of extracellular RNase in *Neurospora crassa*.

3. PMase (Opt. pH 6) activity in the culture in Medium A appeared only in the stationary phase of growth. In Medium B it appeared in the logarithmic phase and the activity was much higher than that in the stationary phase in Medium A. The absence of inorganic phosphate or the presence of RNA seems to stimulate the formation or release of extracellular PMase.

4. PDase (Opt. pH 3) activity in the culture medium in the logarithmic phase increased in parallel with decrease of inorganic phosphate in the medium. The absence of inorganic phosphate seems to stimulate the release of extracellular PDase.

5. These findings suggest that extracellular PDase and PMase, but not RNase $N_1$, play direct rôles in the growth of the mold.

6. A cell extract of strain 74A-T28-M4 cultured in Medium D was prepared by destroying the cells in a French press. The extract contained PDase and at least two RNases which were named RNase $N_2$ and RNase $N_3$. The extract contained no RNase $N_1$. RNase $N_2$ is a RNase $T_2$-like RNase with no absolute base specificity (EC 2.7.7.17) RNase $N_3$ is similar to RNase $N_1$, but has a larger molecular size.

The activity of endocellular RNases in the extract was highest after 63 h culture and that of endocellular PDase was highest in the aged (187 h) culture.

*Purification:* Partial purification of RNase $N_1$ was carried out as follows: *Neurospora crassa*, strain 74A-T32-M12 was grown with aeration at 30° for 4 days in 5.5 l of minimum medium supplemented with 10 µg/ml adenine. The culture was filtered and the filtrate (specific activity 2) was brought to 80% saturation of ethanol by addition of 99.5% ethanol. The precipitate formed was collected by centrifugation, dissolved in about 30 ml of 0.05 $M$ Tris buffer, pH 7.2 (specific activity 33), and dialyzed overnight against distilled water at 4°. The dialyzed enzyme solution was adjusted to pH 5 and applied to a column of DEAE-cellulose equilibrated with 0.005 $M$ $Na_2HPO_4$. The column was washed with 0.005 $M$ $Na_2HPO_4$. The non-adsorbed fraction (specific activity 110) with enzymatic activity was lyophilized, dissolved in a small amount of water, and dialyzed overnight against

0.005 $M$ Na$_2$HPO$_4$. It was then applied again to a column of DEAE-cellulose equilibrated with 0.005 $M$ Na$_2$HPO$_4$. After washing the column with 25 ml of 0.005 $M$
Na$_2$HPO$_4$, non-linear gradient elution was started. The mixing flask contained 270 ml
of 0.005 $M$ Na$_2$HPO$_4$ and the reservoir contained 500 ml of 0.12 $M$ NaCl—0.12 $M$
NaH$_2$PO$_4$. The active fraction (specific activity 1520) eluted at about 0.02 $M$ NaCl
concentration was lyophilized, dissolved in a small amount of water and dialyzed
against distilled water at 4°. Then it was adjusted to pH 4.5 with 0.2 $M$ ammonium
acetate and loaded on a column of CM-cellulose equilibrated with 0.005 $M$ ammonium
acetate-acetic acid (pH 4.5). After washing with 25 ml of 0.005 $M$ ammonium
acetate-acetic acid (pH 4.5), non-linear gradient elution was started. The flask contained 200 ml of 0.005 $M$ ammonium acetate-acetic acid (pH 4.5) and the reservoir
contained 500 ml of 0.1 $M$ ammonium carbonate (pH 7.8). The active fractions
eluted with 0.02 $M$ ammonium carbonate were collected and lyophilized. The
product was dissolved in a small amount of 0.01 $M$ citrate buffer (pH 6.2), and dialyzed overnight against 0.01 $M$ citrate buffer at 4°. The enzyme solution was lyophilized
and dissolved in 1.5 ml of 0.01 $M$ buffer (specific activity 2070) and applied to a column
of Sephadex G-75 (2.0 cm $\times$ 74 cm). It was eluted with 0.01 $M$ citrate buffer (pH 6.2).
Fractions of 2 ml were collected and assayed for ribonuclease activity and for protein
content and the highest specific activity obtained was 10,400. This was about 5,000
times more than that of the culture filtrate. It is remarkable that although it is only
partially purified the enzyme preparation has even higher activity than homogeneous
ribonuclease T$_1$.

The partial purification of RNases N$_2$ and N$_3$ is carried out as follows: *N. crassa*,
strain 74AY152-M7, Y30539 y(ad$^-$) is grown with aeration in Fries medium supplemented with RNA at 30° for 5 days. The culture is filtered and suspended in liquid
at pH 7.2. The suspension is subjected to treatment in a Waring-blender and French
press, and centrifuged. In this way RNase is extracted from the cells. The supernatant
is brought to 75% saturation of ethanol by addition of 99.5% ethanol. The precipitate is dissolved in water (pH 6.2) and dialyzed. It is then subjected to DEAE-
cellulose column chromatography. Two peaks with RNase activity are separated and
these fractions are further purified by Sephadex-G-100 gel filtration: The RNase
with the larger molecular size is RNase N$_2$ and that with the smaller molecular size
is RNase N$_3$.

The properties of RNases N$_1$, N$_2$, and N$_3$ may be characterized as follows:

| Individual name | EC number | Systematic name | Trivial name | Characters |
|---|---|---|---|---|
| RNase N$_1$ | 2.7.7.26 | Ribonucleate guaninenucleotido-2'-transferase (cyclizing) | Guanyloribo-nuclease | tra, endo, $>$ p $\rightarrow$ Xp, G-s, extra. |
| RNase N$_2$ | 2.7.7.17 | Ribonucleate nucleotido-2'-transferase (cyclizing) | Ribonuclease | tra, endo, $>$ p $\rightarrow$ Xp, Non-s, intra. |
| RNase N$_3$ | 2.7.7.26 | Ribonucleate guaninenucleotido-transferase (cyclizing) | Guanyloribo-nuclease | tra, endo, $>$ p $\rightarrow$ Xp, G-s, intra. |

*Molecular size:* The approximate molecular sizes of RNases $N_1$, $N_2$ and $N_3$ are estimated by gel filtration through Sephadex G-100 and Sephadex G-75 with RNase $T_1$ and RNase $T_2$ as standards. Results suggest that RNase $N_1$ = RNase $T_1$ (M.W. 11,000) $<$ RNase $N_3$ $\ll$ RNase $T_2$ (M.W. 36,000) $<$ RNase $N_2$.

*Optimum pH:*

| | |
|---|---|
| RNase $N_1$ | pH 7.0 |
| RNase $N_2$ | pH 8.0 |
| RNase $N_3$ | pH 6.0 to 7.0 |

*Effects of inhibitors and activators:* The effects of metal ions and other effectors on RNase $N_1$ are shown in Table IV-16. This enzyme is strongly inhibited by $Hg^{2+}$; $Zn^{2+}$, which strongly inhibits RNase $T_1$, does not inhibit RNase $N_1$. The apparent activation by $FeSO_4$ and by $ICH_2COOH$ is worth further investigation.

Table IV-16. *Activators and inhibitors of RNase $N_1$*

| Reagent | Final conc. ($-\log M$) | Remaining activity (%) |
|---|---|---|
| NaCl | 0 | 9 |
| | 1 | 66 |
| | 2 | 81 |
| NaF | 1 | 100 |
| $NaN_3$ | 2 | 100 |
| $AgNO_3$ | 3 | 51 |
| $MgCl_2$ | 1 | 19 |
| | 2 | 67 |
| $CaCl_2$ | 2 | 114 |
| | 3 | 100 |
| $HgCl_2$ | 3 | 20 |
| | 4 | 32 |
| $MnSO_4$ | 2 | 103 |
| | 3 | 62 |
| $ZnSO_4$ | 3 | 108 |
| | 4 | 95 |
| $CuSO_4$ | 3 | 47 |
| $FeSO_4$ | 2 | 245 |
| | 3 | 160 |
| $Fe^{3+}$ | 3 | 92 |
| $ICH_2COOH$ | 2 | 190 |
| | 3 | 115 |
| Histidine | 2 | 106 |
| EDTA | 2 | 110 |

The effects of these effectors on RNase $N_2$ and RNase $N_3$ have not been fully investigated. However, so far it has been found that EDTA, which is not inhibitory to RNases $T_1$, $N_1$ and $N_3$, is a strong inhibitor for RNase $N_2$.

*Stability:* RNase $N_1$ is quite stable in neutral or acid media at 37°. It is resistant to heating at 80 °C for 2 min in acidic media (pH 2—4)I. t is fairly unstable in alkaline medium (pH 9).

*Specificity:* 3'-terminal analysis of the digestion products by RNase $N_1$, $N_2$ and $N_3$, showed that RNase $N_1$ and RNase $N_3$ are specific for the 3'-guanylic acid bond and

RNase $N_2$ has no absolute base specificity. Thus RNase $N_1$ and RNase $N_3$ seem to be $T_1$-like enzymes, and RNase $N_2$ is a $T_2$-like enzyme.

When RNA digestion is estimated by the increase of acid-soluble nucleotide, RNase $N_1$ seems to have even higher specific activity than RNase $T_1$. However, it was found that it is far less active in hydrolyzing guanosine 2′,3′-cyclic phosphate to 3′ phosphate than RNase $T_1$. One unit (measured by the increase of acid-soluble nucleotide) of RNase $N_1$, scarcely hydrolyzed G > p. Ten units of RNase $T_1$ hydrolyzed 60% of G > p to Gp in 6 h (Exp. Conditions: G > p, 0.6 μmole: buffer pH 7.0; total volume 0.1 ml; 37°), while comparable hydrolysis by RNase $N_1$ was produced only with 90 units of enzyme in 6 h. RNase $N_3$ is similar in this respect.

RNase $N_2$, an intracellular enzyme, might be the same enzyme as that described by SUSKIND and BONNER, although slight differences, such as in the pH optimum and activation of the latter by phosphate, have been reported.

## F. RNases of Azotobacter agilis (vinelandii)

A nuclease attacking both DNA and RNA has been found in *Azotobacter agilis* by A. SEVENES and R. J. HILMOR. An enzyme specific to RNA has been described recently by SHIIO, ISHII, and SHIMIZU (1966). This RNase is remarkable in two aspects: it is located in particles sedimentable at 105,000 × g and it digests RNA completely to four nucleoside 2′,3′-cyclic phosphates, without further hydrolysis to 3′-nucleotides.

*Preparation:* Centrifugal fractionation of enzyme extract

*Azotobacter agilis* (strain 0) is grown in Burk's nitrogen-free medium. The cells are grown at 30 °C, collected, washed, and kept overnight at 2 °C. Washed cells are disrupted by osmotic shock. The shockate is fractionally centrifuged and 84% of the total RNase activity is recovered in the 105,000 × g pellet fraction.

Localization of RNase in the 105,000 × g particles is observed when the shockate is prepared either in the presence or in the absence of MgSO$_4$ and when the bacterial cells are disrupted by sonic oscillation or by EDTA lysozyme treatment. These findings suggest that the RNase is located in 105,000 × g particles in living cells and the enzyme is firmly bound to the particles.

In addition to RNase, acid phosphatase is observed exclusively in 105,000 × g particles, while alkaline phosphatase is found mainly in the 105,000 × g supernatant, and phosphodiesterase in the 10,000 × g pellet. Polynucleotide phosphorylase is found both in 105,000 × g particles and in the 105,000 × g supernatant. The latter fraction contains 34% of the polynucleotide phosphorylase. The localization of these enzymes participating in phosphate metabolism should be explained in relation to their physiological rôles.

*Solubilization and purification:* When 105,000 × g particles suspended in 20 mM Tris-HCl buffer, pH 7.4, are incubated with EDTA for 4 h at 5 °C and then dialyzed to remove EDTA, the RNase of the particles is solubilized and recovered almost completely in the 105,000 × g supernatant fraction. The supernatant is subjected to DEAE-cellulose column chromatography and the fraction with the highest specific activity is 60 to 70 times as active as the shockate.

*Properties: Stability.* The purified enzyme in an ice bath at pH 7.4 does not lose its activity for a week after preparation but later the activity decreases gradually. It is more stable in neutral and alkaline pH values (pH 10) than at low pH values (pH 3). Excessive dialysis decreases the activity.

Like other RNases it is fairly beat-stable in acid medium. Heating the enzyme solution at 80 °C for 15 min at pH 3.1 scarcely affects the activity. It is less resistant to heating at neutrality.

*Optimum pH.* The optimum pH of the enzyme is about 7.5 in the presence of the activator, $CaCl_2$, but is 7.0 in the absence of $CaCl_2$.

The effects of metallic ions and other effectors are shown in Table IV-17. It requires $Ca^{2+}$ and is inhibited by $Cu^{2+}$ and $Zn^{-+}$.

Table IV-17. *Effect of metallic salts and other substances on the activity of RNase*

| Addition | Concentration (m$M$) | Percent of activity (%) |
|---|---|---|
| None | — | 100 |
| $CaCl_2$ | 0.01 | 113 |
| | 0.10 | 162 |
| | 1.00 | 213 |
| | 10.0 | 227 |
| $MgCl_2$ | 0.10 | 100 |
| | 1.00 | 125 |
| | 10.0 | 125 |
| $MnCl_2$ | 0.10 | 100 |
| | 1.00 | 87 |
| | 10.0 | 30 |
| $CuSO_4$ | 1.00 | 22 |
| $ZnCl_2$ | 1.00 | 17 |
| EDTA | 1.00 | 82 |
| | 10.0 | 54 |
| PCMB | 0.5 | 85 |
| Urea | 1.0 $M$ | 41 |

RNase activity was assayed under standard conditions. The reaction mixture contained 3 mg of commercial yeast RNA, 50 m$M$ Tris-Cl, pH 7.4, about one unit of enzyme, and the given addition in a final volume of 1.0 ml.

*Specificity.* Both highly polymerized yeast RNA and the RNase A-core are completely degraded by *A. agilis* RNase to four nucleoside 2′,3′-cyclic phosphates (A > p, C > p, U > p and G > p). No spot corresponding to Xp could be detected by paper chromatography of the reaction products. The rates of formation of nucleoside cyclic phosphates by the enzyme reaction are in the following order: A > p; G > p; C > p; U > p. These 2′,3′-cyclic nucleotides cannot be hydrolyzed by the enzyme.

Poly A, poly C, poly U and poly I are degraded by the enzyme producing the corresponding nucleoside 2′,3′-cyclic phosphates. Except with poly I, oligonucleotides are identified as intermediates. So the enzyme may be regarded as an endoribo-

nuclease. The rate of degradation of these homopolynucleotides by the enzyme is in the following order:
poly A; poly C; poly U; poly I

Summarizing, the RNase may be characterized as follows:

| Individual name | EC number | Systematic name | Trivial name | Characters |
|---|---|---|---|---|
| Ribonuclease *Azotobacter agilis* | 2.7.7.17 | Ribonucleate nucleotido-2′-transferase (cyclizing) | Ribonuclease | tra, endo, $>$ p, Non-s, intra, part |

## G. RNase I of E. coli

An RNase, which is now designated as RNase I of *E. coli*, has been purified from *E. coli* ribosomes by urea treatment, ammonium sulfate fractionation and chromatography on amberlite CG-50, by SPAHR and HOLLINGWORTH (1961). The purified preparation is free of phosphatase and deoxyribonuclease, but the specific activity is about one tenth of that of pancreatic RNase.

*Properties of RNase I of E. coli* (SPAHR, 1966):

*Optimum pH*. pH 8.1

*Heat stability*. It is rather heat stable in acidic medium. It retains half its activity when heated for 10 min at 100° at pH 3.1.

*Effect of NaCl, $Mg^{2+}$ and sodium dodecyl sulfate*. RNase I activity is greatly enhanced by the presence of monovalent cations such as $K^+$ and $Na^+$. In the presence of 0.2 $M$ NaCl or KCl the activity is about 1.6 times higher than that assayed in 0.1 $M$ Tris buffer alone. It does not require a divalent cation for its activity. $Mg^{2+}$ is rather inhibitory. It is fully active in a concentration of EDTA as higher as 0.05 $M$. Sodium dodecyl sulfate is highly inhibitory, probably because it denatures the enzyme.

*Mechanism of action and specificity:* *E. coli* RNase I cleaves all internucleotide bonds in RNA producing first nucleoside 2′,3′-cyclic phosphates. The cyclic phosphate bond is then hydrolyzed to produce exclusively nucleoside 3′-phosphates. It hydrolyzes A $>$ p and C $>$ p (6-aminonucleotides) 5 times faster than G $>$ p and U $>$ p (6-keto nucleotides). In the initial stage of RNA digestion the enzyme releases more adenylic and uridylic acids than guanylic and cytidylic acids.

It may be regarded as a ribonucleate nucleotido-2′-transferase (cyclizing).

As is well known, pancreatic RNase (RNase I-A) is a basic protein. From the chromatographic behavior, RNase I of *E. coli* seems to be even more basic than RNase I-A.

*RNase from the debris of E. coli:* ANRAKU and MIZUNO (1965, 1967) found that the cell debris of *E. coli*. B contains a considerable amount of RNase, which is different from *E. coli* RNase I.

*E. coli*. B cells were ground with acid-washed sea sand and the ground cell material was extracted with buffer solution, centrifuged and washed successively at 5,000 $\times$ $g$ for 5 min and 20,000 $\times$ $g$ for 20 min giving a "debris" fraction. The RNase in the debris fraction was purified 12,500 fold by treatment with 4 $M$ urea and 0.05 $M$ NaCl, ammonium sulfate fractionation at pH 4.5 and chromatography on CG-50 resin (from Organo Chemical Company, Japan).

In this way ANRAKU and MIZUNO prepared debris-RNase (d-RNase) and at the same time ribosome RNase (r-RNase, probably identical with RNase I) and compared the properties of the two enzymes.

Optimum pH is pH 7.4 to 7.8 for both RNases. The activities of both RNases at pH 7.6 in phosphate buffer are twice as high as those in Tris-HCl.

The effects of several cations on the two RNases are shown in Table IV-18. Both enzymes behave in essentially the same way with these cations.

The effects of urea on the two enzymes are the same. The maximum inhibition by urea is observed with 2 *M* urea.

Qualitatively NaCl was found to activate both RNases as described by SPAHR and HOLLINGSWORTH for RNase I, although quantitatively its effects were somewhat different. Indeed, NaCl activates r-RNase and the d-RNase of ANRAKU and MIZUNO at concentrations of up to only 0.1 *M* while at higher concentrations it is inhibitory.

r-RNase and d-RNase may be distinguished by the difference in their heat stabili-

Table IV-18. *Effect of divalent cations on en-*
*zyme activity*

| Effector | Relative activity | |
|---|---|---|
| | r-RNase | d-RNase |
| None | 100 | 100 |
| MgCl$_2$ 0.01 M | 118 | 132 |
| MnCl$_2$ 0.01 M | 43 | 61 |
| CoCl$_2$ 0.01 M | 29 | 44 |
| CaCl$_2$ 0.01 M | 155 | 142 |
| BaCl$_2$ 0.01 M | 154 | 134 |

ties. The former is less heat-stable than the latter. Thus, the latter retains 76% of its activity after heat treatment at 80 °C for 5 min at pH 6.6 (phosphate buffer) while the former retains only 47% of its activity after the same treatment.

Both enzymes may be further distinguished by their behavior against various RNA's, ribosomal RNA, soluble RNA and commercial low molecular (about 30 nucleotide units on the average) yeast RNA. r-RNase seems to have more tendency to hydrolyze molecules of RNA of high molecular weight than d-RNase, which acts readily on RNA molecules of low molecular weight.

Both enzymes act well on s-RNA, and even more readily than on low molecular yeast RNA. This is rather remarkable, for most RNases attack the latter more readily than the former, which is considered to contain a large amount of double stranded structure.

Both enzymes release adenylic acid faster than guanylic acid as a dialyzable digestion product from RNA and a core with a high content of guanine accumulates in the later stage of digestion. In this respect no remarkable difference can be observed between the two enzymes.

d-RNase can be clearly distinguished from r-RNase by its chromatographic behavior. However, the two enzymes are so similar that it is rather natural to consider that they are slightly modified forms of a single enzyme protein. This relation might be compared to the relation between RNase I-A and RNase I-B (two forms of pancreatic RNase) or RNase $T_2$-A and RNase $T_2$-B. Anyhow it is rather premature to discuss the relation between d-RNase and r-RNase before studying the chemical natures of these enzymes.

## H. RNase of Aspergillus saitoi

M. IRIE isolated an RNase (RNase $M$) from a digestive "Molsin" (*Aspergillus saitoi*) (1967). It was purified about 125-fold by column chromatography using IRC-50, phospho-cellulose, DEAE-cellulose and Sephadex G-75.

Table IV-19. *Effects of divalent cations, EDTA and PCMB on A. saitoi RNase*

| Inhibitor | Remaining activity Inhibitor concentration | |
|---|---|---|
| | $10^{-3}M$ | $10^{-4}M$ |
| None | 100 | 100 |
| $Mg^{2+}$ | 112 | |
| $Ca^{2+}$ | 116 | |
| $Mn^{2+}$ | 108 | |
| $Fe^{2+}$ | 98 | |
| $Co^{2+}$ | 91 | 104 |
| $Ni^{2+}$ | 83.5 | |
| $Cu^{2+}$ | 10.5 | 15.5 |
| $Zn^{2+}$ | 14.3 | 36.8 |
| $Cd^{2+}$ | 42.6 | 89 |
| $Ba^{2+}$ | 116 | |
| $Hg^{2+}$ | | 64.5 |
| $Pb^{2+}$ | 73.0 | 112 |
| PCMB | 45.0 | 91.5 |
| EDTA | 106 | |

Reaction conditions: A 1 ml volume of reaction mixture contained 4.6 µg enzyme, 2.5 mg. RNA and $10^{-3}$ or $10^{-4}M$ of inhibitor, (buffer, 0.025 $M$ acetate, pH 5.5). After 5 min incubation at 37°, the reaction was stopped by the addition of 0.5 ml of MacFadyen reagent and centrifuged. The supernatant was diluted 15 fold with water and the absorbance at 260 mµ was measured. Figures in the table indicate the percentage of activity remaining.

*Properties of the purified RNase M:* It has no DNase activity and no phosphomonoesterase activity with $p$-nitrophenyl phosphate as a substrate. No phosphodiesterase activity with bis-($p$-nitrophenyl) phosphate could be detected.

The sedimentation constant is 3.22, so the molecular size may be similar to that of RNase $T_2$.

It is resistant to heat treatment at 70° for 5 min at pH 5 — 7. However, it is markedly inactivated by heating at pH 3.5 — 4.0. At increasing alkaline pH values, stability decreases gradually. Generally speaking it seems to be less heatstable than small-molecular RNases such as RNase $T_1$.

The optimum pH for the digestion of RNA and for the hydrolysis of U > p is pH 4.0. The effect of ionic strength (with NaCl) on RNase $M$ activity was studied using RNA as substrate at pH 4.0. So far as tested, the activity is found to increase with increase of ionic strength. The activity at an ionic strength of 1.0 is 125% of that at an ionic strength of 0.05. The effects of various effectors are summarized in Table IV-19.

The inhibition by PCMB is rather exceptional for an RNase. The enzyme requires no divalent cations.

*Substrate specificity*: A > p, G > p, C > p, U > p are all hydrolyzed to 3'-nucleotides. The rates of the hydrolysis are in the order: G > p; U > p; C > p; A > p.

During the early stage of digestion of RNA by the enzyme, the release of Ap was pronounced and subsequently increases of C > p, Up and Gp were observed.

In this respect, it is very similar to RNase $T_2$. So its characteristics may be summarized as follows:

| Individual name | EC number | Systematic name | Trivial name | Characters |
|---|---|---|---|---|
| RNase M | 2.7.7.17 | Ribonucleate nucleotido-2'-transferase (cyclizing) | Ribonuclease | tra, endo, > p → Xp, Non-s |

# I. RNase of Bacillus subtilis

Extensive studies on the purification, specificities, and chemical nature of extracellular ribonucleases have been carried out by S. NISHIMURA and his coworkers (NISHIMURA and NOMURA, 1959; NISHIMURA, 1966). The specificity of one of these enzymes has been further investigated by WHITFELD and WITZEL (1963) and by RUSHITZKY et al. (1963). Another extracellular RNase has been described by NIKAI et al. (1965). *Bac. subtilis* produces an intracellular RNase, which is quite different from the extracellular RNases.

## 1. Extracellular RNases of Bacillus subtilis Strain H

(NISHIMURA and NOMURA, 1959; NISHIMURA, 1966).

Two forms of extracellular RNase were isolated from the culture medium of *Bacillus subtilis* strain H by column chromatography on IRC-50 or CM-cellulose. They do not differ from each other with respect to the effects of EDTA and metal ions, or in their enzymatic specificities on RNA, so only the main component has been extensively investigated.

The main component was purified from the culture medium by acid treatment, adsorption on IRC-50, ammonium sulfate fractionation and finally by CM-cellulose column chromatography. In this way the extracellular RNase can be obtained in a homogeneous state. For its crystallization, the enzyme preparation is dissolved in distilled water at a concentration equivalent to 4 mg of tyrosine/ml. The enzyme is gradually crystallized from the solution by adding saturated ammonium sulfate solution in the cold.

*Properties of the extracellular RNase of Bac. subtilis:* The purified preparation has no DNase, PDase or PMase activity. It is homogeneous on chromatography and ultra-centrifugation. It has a single N-terminal amino acid (Serine).

Optimum $pH$ 7.5   (Nishimura; phosphate buffer)
           8.5   (Rushizky et al.: Tris-HCl buffer).

The effects of metallic ions and other effectors are shown below (Table IV-20): (Nishimura, 1960).

Table IV-20. *Effectors of Bacillus subtilis RNase*

| Inhibitor | Conc. | Percent of activity remaining |
|-----------|-------|-------------------------------|
| PCMB | $1.0 \times 10^{-2} M$ | 100 |
| $Pb^{2+}$ | $1.0 \times 10^{-2} M$ | 91 |
| $Mg^{2+}$ | $1.0 \times 10^{-2} M$ | 91 |
| $Fe^{2+}$ | $1.0 \times 10^{-2} M$ | 77 |
| $Fe^{3+}$ | $1.0 \times 10^{-2} M$ | 65 |
| $Ca^{2+}$ | $1.0 \times 10^{-2} M$ | 85 |
| $Mn^{2+}$ | $1.0 \times 10^{-2} M$ | 85 |
| $Zn^{2+}$ | $1.0 \times 10^{-2} M$ | 65 |
| $Cu^{2+}$ | $1.0 \times 10^{-2} M$ | 71 |
| $Co^{2+}$ | $1.0 \times 10^{-2} M$ | 61 |
| $Hg^{2+}$ | $1.0 \times 10^{-2} M$ | 14 |
| EDTA | $1.7 \times 10^{-2} M$ | 100 |
| Polyvinyl sulfate | 500 µg/mg | 114 |

EDTA has neither an activating nor an inhibitory effect on the enzyme. It is noteworthy that metal ions, such as $Pb^{2+}$, $Cu^{2+}$ and $Zn^{2+}$, have no remarkable inhibitory effect on the enzyme. According to Rushizky et al., $CuSO_4$ and $MgSO_4$ cause 96% and 81% inhibition, respectively, at the high concentration of 0.1 $M$.

The sedimentation coefficient of the enzyme was found to be 1.4 by Nishimura (1960). Hartley et al. (1963) estimated the molecular weight to be 10,700 $\pm$ 400 from the partial specific volume ($v = 0.703$) and sedimentation coefficient ($S°_{20, w} = 1.55$).

*Amino acid composition:* Nishimura and Ozawa (1962) analyzed the amino acid composition of the enzyme (Table IV-21).

The most remarkable characteristic of this is the absence of cysteine and therefore the absence of disulfide bridges. Pollock and Richmond (1962) suggest that the absence of disulfide bridges might be a general characteristic of bacterial exo-enzymes. They suggested that bacterial exo-enzymes might thus have more than the usual degree of flexibility and be more capable of unfolding and refolding than proteins which contain disulfide bridges. The possibility that these bacterial enzymes may have become adapted so as to pass more freely through the rigid cell-wall meshwork by evolving a degree of flexibility through the elimination of disulfide bridges seems at least worthy of consideration.

Unlike Nishimura and Ozawa, Hartley et al. (1963) reported that the enzyme contains no methionine residue. It contains three histidine residues per molecule just like RNase $T_1$. It is inactivated by photooxidation in the presence of methylene blue due to the specific degradation of the histidine residues or by the action with N-bromosuccinimide due to the selective cleavage of C-tryptophyl peptide bonds (NISHIMURA and OZAWA). These findings, although quite indirectly, suggest the possibility that at least one of the histidine residues and one of the tryptophan residues may participate in the active center of the enzyme.

*Specificity. Bac. subtilis* extracellular RNase splits the phosphodiester bonds of the nucleoside 3'-phosphates in RNA. 2',3'-cyclic nucleotides are formed as intermediates, and these are further hydrolyzed to form 3'-nucleotides. Formerly it was considered that the enzyme was a purine specific RNase (NISHIMURA). However,

Table IV-21. *Amino acid composition of Bacillus subtilis RNase*

|  | Moles of amino acid residue per mole of *Bacillus subtilis* RNase |
| --- | --- |
| Asp | 17.4 |
| Thr | 8.8 |
| Ser | 9.3 |
| Glu | 7.8 |
| Pro | 5.3 |
| Gly | 10.8 |
| Ala | 8.0 |
| $^{1}/_{2}$ Cys | 0 |
| Val | 4.3 |
| Met | 0.7 |
| Ile | 7.1 |
| Leu | 7.1 |
| Tyr | 7.5 |
| Phe | 5.9 |
| Lys | 8.5 |
| His | 3.0 |
| Arg | 6.1 |
| Trp | 4.0 |
| NH₃ | (14.8) |
| | 121.6 |

extensive studies on the specificity of the enzyme by WHITFELD and WITZEL (1963) and by RUSHIZKY et al. (1963) have revealed that its specificity was far more complex.

Generalizations on the relative base specificity of the enzyme are very difficult because of the effects of adjacent nucleotides. However, with a few exceptions, the phosphodiester bonds of 3'-purine nucleotides are cleaved faster than those of 3'-pyrimidine nucleotides, and those of 3'-nucleotides with a 6-oxo group are cleaved faster than those of 3'-nucleotides with a 6-amino group. So when adjacent bonds are the same, the following relations may be pointed out:

— GpNp — > — ApNp —
— UpNp — > — CpNp —
— GpNp — > — UpNp —
— ApNp — > — CpNp —
— GpNp — > — CpNp —

The enzyme cleaves — GpGp — and — GpAp — bonds about 100 times faster than other phosphodiester bonds in RNA (RUSHIZKY et al.). According to RU-SHIZKY et al., another characteristic of the enzyme is the resistance of Gp-terminal dinucleotides to hydrolysis. So the oligonucleotides produced by digestion of RNA with RNase T$_1$ will give the dinucleotide core NpGp as digestion products: ApApCp UpGp → UpGp and other digestion products.

WHITFELD and WITZEL compared the susceptibilities of various dinucleotides and dinucleoside monophosphates and they obtained the following series:

$$GpGp > GpA > GpC > ApA > p > ApAp > ApA.$$

The last compound is scarcely cleaved by the enzyme. Thus, in general, GpNp > IpNp > ApNp > UpNp > CpNp and NpNp > NpN.

The order of hydrolysis of nucleoside 2',3'-cyclic phosphates is as follows:

$$G > p > I > p > A > p, C > p, U > p.$$

WHITFELD and WITZEL studied the effects of the enzyme on the homopolymers, poly I, poly A, poly U and poly C. They found the following order of susceptibility: poly I > poly A > poly U > poly C. The last compound is scarcely attacked. They did not examine the action on poly G but suggested that it would be degraded most easily. However, as poly G has a high tendency to aggregate, the attitude of poly G to the enzyme cannot be so simply deduced.

WHITFELD and WITZEL have shown the degradation pattern of poly A as follows:

The products of exhaustive digestion of RNA by the enzyme are:
guanylic acid   (mainly G > p);
adenylic acid, uridylic acid, and cytidylic acid (mainly A > p, U > p, C > p) and di- and trinucleotides mainly with the terminals, Gp, A > p, U > p and C > p.

## 2. Extracellular RNase of Bacillus subtilis Marburg Strain

A quite different RNase was isolated by NIKAI et al. (1965) from the culture medium of Bac. subtilis Marburg strain. Its optimum pH is 5.0.

It degrades RNA to form mainly A > p, G > p, U > p and C > p. Only small amounts of 3'-nucleotides are detected as digestion products. The enzyme is heat labile and inhibited by EDTA. Ca$^{2+}$ stabilizes the enzyme.

## 3. Intracellular RNase of Bacillus subtilis

NISHIMURA and MARUO (1960) extracted an RNase from cells of Bac. subtilis strain H. It is quite different from the extracellular RNases of the same strain. Thus, it does not react with the antibody for the extracellular enzyme.

*Properties:* The optimum pH is 5.8 and the enzyme is inhibited completely by $1.7 \times 10^{-3}M$ EDTA. The digestion products of RNA are $A > p, G > p, C > p$ and $U > p$.

The RNases of *Bac. subtilis* may be summarized as follows:

| Individual name | EC number | Systematic name | Trivial name | Characters |
|---|---|---|---|---|
| RNase, *Bac. subtilis* H | 2.7.7.17 | Ribonucleate nucleotido-2-transferase (cyclizing) | Ribonuclease | tra, endo, $> p(\rightarrow Xp)$, Non-s (relative specificity difficult to define), extra |
| RNase, *Bac. subtilis* Marburg | 2.7.7.17 | Ribonucleate nucleotido-2-transferase (cyclizing) | Ribonuclease | tra, endo, $> p(\rightarrow Xp)$, Non-s, extra |
| RNase, *Bac. subtilis* (intracellular) | 2.7.7.17 | Ribonucleate nucleotido-2-transferase (cyclizing) | Ribonuclease | tra, endo, $> p$, Non-s, intra |

## J. Yeast RNase

Y. OHTAKA et al. (1963) have purified an intracellular RNase from the autolysate of pressed baker's yeast (Oriental Yeast Co). It is a novel type of RNase, differing from other RNases as regards its mode of action on RNA.

*Properties:* The optimum pH is 7.3 (Tris buffer). It is heat sensitive like most common enzymes. Above 45° it is labile, and completely destroyed by heating at 65° for 10 min. The effects of metal ions and other effectors are shown in Table IV-22.

Table IV-22. *Activation and inhibition*

| | Inhibition (%) | | | Activation (%) | |
|---|---|---|---|---|---|
| $ZnSO_4$ | $10^{-4}M$ | 100 | $KH_2PO_4$ | $10^{-1}M$ | 80 |
| | $10^{-5}M$ | 90 | | $10^{-2}M$ | 40 |
| $CaCl_2$ | $10^{-2}M$ | 100 | NaCl | $10^{-2}M$ | 15 |
| | $10^{-3}M$ | 60 | | | |
| $CoCl_2$ | $10^{-2}M$ | 100 | KCl | $10^{-2}M$ | 10 |
| | $10^{-4}M$ | 60 | | | |
| $CuSO_4$ | $10^{-3}M$ | 50 | PCMB | $10^{-3}M$ | 0 |
| $MnSO_4$ | $10^{-3}M$ | 20 | EDTA | $10^{-3}M$ | 20 |
| $CaCl_2$ | $10^{-4}M$ | 15 | | | |
| $MgCl_2$ | $10^{-2}M$ | 0 | | | |

Although the enzymatic reaction proceeds in the absence of phosphate, remarkable activation was observed on its addition. The enzyme was strongly inhibited by various metal ions, especially $Zn^{++}$. The inhibition by $Zn^{++}$ can be reversed by EDTA.

*Specificity:* The enzyme hydrolyzed RNA completely to acid-soluble products, which were separated on a Dowex-1 column and confirmed to be Ap, Gp, Up and Cp. It splits neither DNA, diphenyl phosphate nor 2′,3′-cyclic nucleotides. So the enzyme may be tentatively classified as ribonucleate 3′-nucleotidohydolase.

| Individual name | EC number | Systematic name | Trivial name | Characters |
|---|---|---|---|---|
| Yeast RNase | — | Ribonucleate 3′-nucleotidohydrolase | Yeast RNase | hyd, exo, → Xp, RNase, Non-s intra. |

## K. RNase II of E. coli

A ribonuclease differing from *E. coli* RNase I has been purified from an extract of *E. coli* strain-B by SPAHR (1964). The enzyme designated as *E. coli* RNase II can be obtained and purified either from the supernatant fraction or the ribosomes of *E. coli*.

*Some properties of E. coli RNase II:*

*Effect of ions*: RNase II requires the presence of both a monovalent cation (K⁺ or NH⁺) and a divalent cation (Mg²⁺ or Mn²⁺) for its activity. Na⁺ cannot replace K⁺ or NH⁺. The optimum concentrations of monovalent and divalent cations are $5 \times 10^{-2} M$ and $10^{-4} — 10^{-3} M$, respectively.

*Effect of pH.* The optimum pH is 7.0 — 8.0.

*Stability.* *E. coli* RNase II is remarkably labile for an RNase. Even when kept in the cold or frozen, it loses 90% of its activity after 24 h. The enzyme can be stabilized by adding bovine serum mercaptalbumin, and in its presence the enzyme can be stored in the lyophilized state without appreciable loss of activity.

*Inhibition.* The activity of RNase II on poly U was assayed in the presence of various substances. The results are shown below:

The enzyme was completely inhibited by 1 $M$ urea; in 0.05 $M$ urea, 11% of the activity was inhibited. RNase II was inhibited by a concentration of EDTA capable of binding all the Mg²⁺ present. If, however, the Mg²⁺ concentration was raised, full activity was restored. Sucrose inhibited RNase II only slightly.

*Mode of action and specificity.* RNase II degrades poly U to form 5′-UMP and very small amounts of oligouridylic acids. This observation suggests that the enzyme acts on poly U predominantly as an exonuclease, but that it can also act partly as an endonuclease. Indeed it degrades RNA to form four 5′-mononucleotides and oligonucleotides. It is practically inactive on DNA.

| Individual name | EC number | Systematic name | Characters |
|---|---|---|---|
| RNase II of *E. coli* | — | Ribonucleate 5′-nucleotidohydrolase | hyd, exo (and endo), → pX RNase, Non-s, intra. |

## L. RNase III of E. coli

An RNase quite recently designated as *E. coli* RNase III (ROBERTSON, WEBSTER, and ZINDER, 1967) is characterized by the activity specifically to digest double-stranded RNA. Double-stranded RNA's such as Reovirus RNA and the double-stranded replicative form of phage R 17 are digested by the enzyme at least ten times more rapidly than single-stranded $f_2$ phage RNA.

It is present in extracts of *E. coli* K 12 prepared for the *in vitro* amino acid incorporating system according to NIRENBERG and MATTHAEI, and can be concentrated and partially purified in a stable form from the extracts.

As most of the RNases (other *E. coli* RNases, RNases $T_1$, $T_2$, $U_2$, $U_3$ etc.) are specific for single-stranded RNA, this RNase will be a valuable tool for the conformation analysis of RNA.

## M. General Considerations

In the foregoing section the RNases, which have been studied in more or less full chemical detail, have been presented. Here they will be surveyed from the viewpoint of comparative biochemistry.

In 1964 RUSHIZKY et al. (1964) purified RNases from the culture media of several microorganisms by essentially similar procedures: acid treatment, ammonium sulfate precipitation, dialysis, and column chromatography. In this way they obtained RNase $T_1$-like G-specific RNases from the culture media of *Bacillus pumilus* IFO 3028 and *Mucor genevensis* IFO 4585, and RNases without strict base specificity from the culture media of *Bacillus cereus* ATCC 10987, *Lenzitis tenuis* IFO 4946 and *Monascus polosus* IFO 4480. They consider that guanyloribonucleases are generally of small molecular size (m.w. 10,000 — 13,000) while nonspecific RNases have molecular weights of about 30,000 — 40,000. However, it seems to us that far more extensive chemical studies will have to be carried out before a general conclusion can be drawn.

Although little information is yet available, we will briefly review the general characteristics of RNases from a comparative biochemical viewpoint (Table IV-23).

### 1. Ribonucleate Guaninenucleotido-2'-transferase (Cyclizing)

The properties of 12 guanyloribonucleases are shown in Table IV-23.

These properties are very similar. The optimum pH is around neutrality or slightly on the alkaline side. These enzymes are generally heat stable and have a molecular weight of about 10,000 or a little more. Neither SH-reagents nor EDTA are inhibitory. $Zn^{2+}$ and $Cu^{2+}$ are generally strongly inhibitory for guanyloribonucleases.

It seems likely that some of these enzymes will be completely purified and then their chemical structure can be compared with that of RNase $T_1$. This will provide important information on the relationship between the structure and function of guanyloribonucleases and for comparative biochemistry including biochemical taxonomy and chemical evolution.

### 2. Ribonucleate Purinenucleotido-2'-transferase (Cyclizing)

It is, of course, still impossible to generalize on the characteristics of puryloribonucleases. So far they seem similar to guanyloribonucleases in molecular size,

Table IV-23. *Properties of microbial ribonucleases*

| Individual name or source of enzymes | pH opt. | Heat-stability | M.W. | Effectors | | | | | | | | |
|---|---|---|---|---|---|---|---|---|---|---|---|---|
| | | | | SH-reagents | EDTA | Ca²⁺ | Mg²⁺ | Zn²⁺ | Cu²⁺ | Hg²⁺ | Ag⁺ | DFP |
| a) Guanyloribonucleases: tra, endo, >p → Xp, G-s | | | | | | | | | | | | |
| RNase T₁ | 7.5 | yes | 11,000 | | | | | | | | | — |
| *Streptomyces erythreus* | 7.3 — 7.4 | yes | | — | — | — | — | I | I | I | I | |
| *Str. albogriseolus* | 7.0 — 8.5 | yes | | — | | | | I | | | | |
| *Actinomyces aureoventicillatus* | 7.6 | | | | | | | | — | | | |
| RNase U₁ | 8.0 — 8.5 | yes | ca. 10,000 | — | — | — | — | I | I | | | |
| *Ustilago zeae* | 8 — 9 | no | | | | | | | | | — | |
| RNase N₁ | 7.0 | yes | ca. 10,000 | sA | — | | I | | I | I | | |
| RNase N₃ | 6 — 7 | yes | a little larger than N₁ | | — | | | — | I | | | |
| *Acrocylindrium* sp. NM 2 | | | | | | | | | | | | |
|   I | 8.0 | yes | | — | — | — | — | I | | | | |
|   II | 8.0 | | | | — | | | I | I | | | |
| *Bacillus pumilus* | 7.9 | | 10,000 — 15,000 | | — | | | | | | | |
| *Mucor genevensis* | 7.9 | | 10,000 — 15,000 | | I | | | | | | | |
| b) Puryloribonucleases: tra, endo, >p → Xp, Pur-s | | | | | | | | | | | | |
| RNase U₂ | 4.5 | | ca. 10,000 | — | — | — | — | — | I | I | | |
| RNase U₃ | 4.5 | | ca. 10,000 | — | — | — | — | — | I | I | | |

Table IV-23 (continued)

c) Ribonucleate nucleotido-2'-transferases (cyclizing): tra, endo $>$ p $\rightarrow$ Xp, Non-s

| | pH | | MW | | | | | | | | Remarks |
|---|---|---|---|---|---|---|---|---|---|---|---|
| RNase T₂ | 4.5 | yes | 36,000 | — | — | I | I | I | I | I | — |
| RNase N₂ | 8.0 | | >36,000 | I | — | — | I | I | I | | |
| RNase I of *E. coli* | 8.1 | yes in acid | | I | I | — | I | I | | | activated by Na⁺, K⁺ |
| RNase M | 4.0 | yes | 30,000 | — | — | I | I | I | | | |
| (*Asp. saitoi*) | | | | | | | | | | | |
| *Bac. subtilis* H | 7.5 — 8.5 | yes | 10,700 | — | I | — | I | I | — | | |
| *Bac. subtilis* | 5.0 | no | | I | I | I | I | I | | | |
| Marburg | | | | | | | | | | | |
| *Bac. subtilis* H | 5.8 | | | | | | | | | | |
| (intra) | | | | | | | | | | | |
| *Bacillus cereus* | 7.9 | | 30,000 | | | | | | | | |
| | | | —40,000 | | | | | | | | |
| *Lengites tenius* | 7.9 | | 30,000 | | | | | | | | |
| | | | —40,000 | | | | | | | | |
| *Azotobacter agilis* | 7.5 | yes in acid | | essential | | | | | | | no detectable hydrolase activity |

d) Other Ribonucleases

| | pH | | | | | | | | | | Remarks |
|---|---|---|---|---|---|---|---|---|---|---|---|
| Baker's yeast (hyd, | 7.3 | no | | — | — | — | I | I | | | activated by phosphate |
| exo, $\rightarrow$ Xp, Non-s) | | | | | | | | | | | |
| RNase II of *E. coli* | 7 — 8 | no | | I | | | | | | | |
| [hyd, exo (and endo), | | | | | | | | | | | |
| $\rightarrow$ pX, Non-s] | | | | | | | | | | | |
| RNase U₄ (hyd, | 8 | | | I | | I | I | | | | |
| exo, $\rightarrow$ Xp, Non-s) | | | | | | | | | | | |

Remarks

—: Neither appreciable activation nor appreciable inhibition.

I: inhibition, sA: slight activation.

A or I denote that appreciable activation or inhibition respectively occurs at the concentration of $10^{-3}$ $M$.

and insensitivity to SH-reagents and EDTA. Unlike guanyloribonucleases, their optimum pH values are on the acidic side.

### 3. Ribonucleate Ribonucleotido-2'-transferase (Cyclizing)

As suggested by RUSHIZKY et al., ribonucleate ribonucleotido-2'-transferases (cyclizing) generally have a molecular weight of 30,000 to 40,000. However, one exception (RNase of *Bac. subtilis* Marburg) should be noted. Most of these enzymes are heat stable like guanyloribonucleases. No generalization can be made on their optimal pH values.

### 4. Other RNases

Ribonucleases, which directly hydrolyze RNA but not DNA, are not included in the Enzyme Commission Report of the IUB. They are rather exceptional enzymes.

In concluding this chapter, we should say that we hope many microbial RNases will be highly purified and their chemical structure will be compared with an interest in comparative biochemistry.

Chapter V

# Physiological Role of RNA-Degrading Enzymes in Microorganisms

## A. Introduction

In the previous chapter, the properties of individual enzymes were discussed mainly from a chemical point of view.

Here, we would like to describe the physiological aspects of RNA-degrading enzymes in microbial cells, for instance, their intracellular location and their rôle in nucleic acid metabolism.

During the last few years, numerous studies on nucleic acids have shown the important rôle of these enzymes in cells. With advances in research on nucleic acids themselves, interest in the metabolism of nucleic acids has increased and consequently various kinds of enzymes participating in the breakdown of nucleic acids have been found in microorganisms and studied. However, no distinct results on the precise rôle of RNA-degrading enzymes in microbial cells have yet been obtained. Therefore, in this monography, the information available at present will be summarized and some working hypotheses based on this information will be considered.

The enzymes that break down RNA in the cells of *Escherichia coli* will be chiefly considered here, because these enzymes have been investigated most thoroughly in this bacteria. Moreover, intensive investigations are being carried out on genetic and biochemical problems of nucleic acid metabolism in *E. coli* which are closely related to the phenomena that will be described in this chapter. These are, for example, the problems of enzymes which catalyze the further degradation of the products formed by RNases, the structure, biosynthesis and biological activities of nucleic acids and genetics at a molecular level. Thus, *E. coli* seems to be the most suitable organism for studies on the general metabolism of nucleic acids and its control in cells. But, as the RNA degrading enzymes found in *E. coli* are all intracellular, the extracellular enzymes in other microorganisms, such as *Bacillus subtilis*, *Neurospora* and *Ustilago*, will also be considered.

## B. Role of Intracellular Enzymes with Special Reference to E. coli

The enzymes in *E. coli* so far known to degrade RNA are RNase I (ribosomal "Latent" ribonuclease), RNase II (potassium activated ribonuclease) and polynucleotide phosphorylase[1]. Polynucleotide phosphorylase was not mentioned in the

---

[1] RNase III, quite recently discovered, is specific for double-stranded RNA. It may participate in some way in the degradation of double-stranded RNA (see Chap. IV, 12).

previous chapter, because it is not strictly an RNase. However, it is known to play an important rôle in RNA metabolism. So, it seems desirable to consider these three enzymes together in relation to cell physiology.

## 1. Mode of Existence

### a) Ribonuclease I (Ribosomal "Latent" Ribonuclease)

This was the first RNA-specific nuclease discovered in *Escherichia coli*. It is bound to ribonucleoprotein particles in cell-free extracts (ELSON, 1958, 1959; BOLTON et al., 1959). The excistence of a ribonuclease bound to the ribosomal fraction in tobacco leaves had previously been reported by RIRIE (1950, 1957). Since its discovery in *E. coli*, ribosomal ribonucleases have been found in yeast (STANESH, 1958), pea seedlings (Tśo et al., 1958), guinea pig pancreas (SIEKEWITZ and PALADÉ, 1958 TASHIRO, 1958). In these materials, however, similar ribonucleases are also present in other fractions of the cells such as the mitochondria and supernatant fractions. In contrast, *E. coli* RNase I is found to be located exclusively in ribosomes in cell-free extracts. The enzyme cannot be removed from the ribosomes by repeated washing, which eliminates most of the nonspecifically adsorbed enzymes, such as $\beta$-galactosidase. This makes it unlikely that the RNase activity is due to contamination. ELSON showed that when ribosomes were prepared carefully in buffer containing 0.01 $M$ $Mg^{2+}$, no enzyme activity was detectable by any method of assay. The activity only appears after the ribosomes have been disrupted into RNA and the protein moiety with agents such as urea or EDTA. Treatment with solutions of high ionic strength (for example, 0.5 $M$ NaCl), tryptic digestion and incubation at 37 °C in 0.1 $M$ NaCl are also effective for the release of the RNase (SPAHR and HOLLINGWORTH, 1961).

From his results, ELSON designated the enzyme as a "ribosomal latent" RNase. It is often called RNase I for simplicity, as it was the first RNase to be found in *E. coli*. The RNase was purified by SPAHR and HOLLINGWORTH (1961) and its properties have been studied in detail, as described in Chapter IV.

The ribosomal nature of the RNase was confirmed by analysis of its behavior on centrifugation. When cell-free extracts are prepared in buffer containing $10^{-2}$ $M$ $Mg^{2+}$ to assure the integrity of 70s ribosomal particles and are subjected to differential centrifugation, all the RNase activity is recovered in the ribosomal fraction. Furthermore, when 70s particles are separated into two subunits, that is the 50s and 30s components, by lowering the $Mg^{2+}$ concentration followed by treatment with urea, RNase activity is found practically exclusively associated with 30s components (SPAHR and HOLLINGWORTH, 1961; ELSON and TAL, 1959; TAL and ELSON, 1963). The slight activity of the 50s fraction can be explained as due to contamination with 30s particles.

The significance of the existence of this RNase in ribosomes is not known. It might be one of the proteins synthesized on ribosomes. In this case, however, some of the activity should be expected to be liberated as soluble RNase. The fact that no RNase I activity can be found in the supernatant fraction may show that this enzyme is not one of the enzymes which is synthesized but that it is an obligatory component which plays some rôle in the metabolism and/or function of ribosomes.

Estimation of the amount of enzyme per mole of ribosome carried out by SPAHR and HOLLINGWORTH (1961) with their purest preparation gave a value of 0.1 mole per mole of 70s particles (molecular weight, $2.8 \times 10^6$), assuming the molecular weight of RNase to be the same as that of pancreatic RNase. Another reported value was between 0.1 and 1 mole/ribosome (BOLTON et al., 1959). The existence of a constant amount of RNase per ribosome seems to support the idea that this enzyme is not a protein nonspecifically bound to ribosomes.

On the other hand, NEU and HEPPEL (1964) discovered that when *E. coli* cells were converted into spheroplasts by the action of EDTA and lysozyme, a ribonuclease was released into the medium as well as some other degradative enzymes. The RNase thus released was purified and its properties were compared with those of ribosomal RNase. As regards their modes of action, pH optima, the effect of $Mg^{2+}$ ions and their heat stabilities, these two enzymes are very similar. The rates of hydrolysis of s-RNA and ribosomal RNA by the two enzymes are especially similar; s-RNA is hydrolyzed about 60% as fast as ribosomal RNA in the presence of $10^{-3}\ M$ EDTA. Although, on DEAE-cellulose column chromatography, a higher concentration of NaCl is required for the elution of the enzyme released into the medium than for that of ribosomal one, this may be due to some minor modification of the enzyme protein caused by different preparation procedures. So, the RNase liberated into the medium during spheroplast formation may simply be "ribosomal RNase". Ribosomes from spheroplasts contain much less RNase than those from intact cells and this deficit is just balanced by the amount of RNase in the medium surrounding the spheroplasts. This gives additional evidence for the identity of the two enzymes. In spite of release of enzyme, gross damage to ribosomal structure is not observed in spheroplasts. Total recovery of ribosomal RNA per gr. of bacterial cells is the same in both intact cells and spheroplasts. Moreover, the patterns obtained by sucrose-density gradient centrifugation of 30s and 50s particles obtained by dissociation of the ribosomes also show that the ratio of protein to RNA is the same in both cases. This means the release of RNA under this condition does not cause the disruption of the ribosomes. This is apparently inconsistent with the *in vitro* observation by ELSON that RNase cannot be released without disruption of ribosomes.

At least two explanations are possible for the apparent contradiction mentioned above. The first explanation is that the *in vitro* observation that all of the RNase I activity in cell-free extracts is bound to 30s particles may be due to the isolation procedure and the bulk of the RNase may indeed be in a free state in cells. This seems probable when it is remembered that, when purified ribosomal RNases are mixed with ribosome suspension, about 10 times more enzyme is adsorbed on to the 30s particles than was originally present on the ribosomes of intact cells (SPAHR and HOLLINGWORTH, 1961). According to this idea, free RNase is concentrated between the cell wall and cell membrane in the logarithmic phase of growth, and so is rapidly released during spheroplast formation. In fact, 90% of the RNase activity is liberated into the medium during their formation. As ribosomes are known to exist in the cytoplasm from the electron micrographs of CONTI and GETTNER (1962), there seems to be a rigid compartment separating the ribosomes and RNase. In the stationary phase, RNase becomes bound to the ribosomes. Experiments show that only about half of the RNase is released when spheroplasts pass into the stationary phase and the other half is found on the ribosomes.

The second explanation is that polyribosomes which are shown clustered at the surface of the bacterial cell (SUIT, 1963; SCHLESSINGER, 1963; HENDLER et al., 1964)

5*

and are active in protein synthesis contain RNase in a somewhat loosely bound form and this RNase is released from polyribosomes by treatment with EDTA to induce spheroplast formation. It is not possible to decide which possibility is the more likely, but anyway the mechanism controlling the activity of RNase in growing cells must be considered. This problem will be discussed later in this chapter.

Two kinds of enzymes which are very similar in their properties to RNase I have been reported. One was found by ANDERSON and CARTER (1965) in the acid soluble fraction when cells were treated with perchloric acid and the other was an enzyme which sedimented with the cell debris on centrifugation of cell extracts (ANRAKU and MIZUNO, 1965, 1967). The former enzyme differs in its behavior on cellulose column chromatography and in the inhibitory effect of NaCl from RNase I purified by SPAHR and HOLLINGWORTH (1961) and the latter also differs from RNase I in chromatographic proporties and in the initial velocity of digestion of various RNAs. RNase prepared from cell debris has a tendency to hydrolyze RNAs of low molecular weight more rapidly than ribosomal RNase. But, in general, these enzymes are rather similar to the ribosomal enzyme. The differences mentioned above might be caused by some changes in structure due to the different preparation procedures or to different components attached to the RNase protein. In fact, the RNase from cell debris, is eluted in the same position as ribosomal RNase on DEAE-cellulose column chromatography after treatment with sufficiently little lysozyme to cause no loss of activity. Since, in a mutant strain which lacks RNase I, the activities of these two enzymes are also undetectable (GASTELAND, 1966), all three enzymes may well be controlled by a single gene, that is originally they were one and the same enzyme.

### b) Ribonuclease II (Potassium Activated Ribonuclease)

In 1961 WADE and LOVETT (1961) reported a phosphodiesterase activity in the ribosomal fraction, which yielded nucleoside 5'-monophosphates as digestion products of RNA and was stimulated by inorganic phosphates. The following year, TISSIERES and WATSON (1962) observed the slow breakdown of messenger RNA in an incubation mixture containing ribosomes, labeled messenger RNA and the supernatant fraction and noticed that the rate of breakdown was greatly increased by adding inorganic phosphates and an ATP generating system to the incubation mixture. They considered that some unknown enzyme participated in this reaction. Subsequently, SEKIGUCHI and COHEN (1963) examined the degradation of phage-induced messenger RNA by polynucleotide phosphorylase in detail and found that, besides polynucleotide phosphorylase, a kind of phosphodiesterase participated in the breakdown of messenger RNA. It was attached to ribosomes and requiring inorganic phosphates for its activity.

SPAHR et al. (SPAHR and SCHLESSINGER, 1963; SPAHR, 1964) showed the existence of a phosphodiesterase which acted only in the presence of $K^+$ (or $NH_4^+$) and $Mg^{2+}$ (or $Mn^{2+}$) ions in cell free extracts. They named this phosphodiesterase "ribonuclease II" because it was discovered second and they considered that the enzymes observed by WADE and LOVETT, TISSIERES and WATSON, and SEKIGUCHI and COHEN, might also be ribonuclease II. SPAHR explained in his report that they misunderstood the $K^+$ ion effect in stimulation of the enzyme as an effect of phosphate or an ATP generating system, because they always added phosphate as the potassium salt and the ATP generating system also contained this cation.

In contrast to RNase I, RNase II is found in both the supernatant and particulate fractions and it is fully active in cell-free extracts (SPAHR, 1964; SINGER and TOLBERT, 1965). The enzymes purified from the two fractions are considered to be identical from observations on properties such as their ion requirement, pH optima, and modes of action.

The modes of attachment of RNases I and II to ribosomes seem to be quite different. A variable amount of RNase II activity is observed in the ribosomal fraction and repeated washing removes much of this activity. After dissociation of the ribosomes into 50s and 30s components, more than 80% of the activity is liberated into the supernatant fraction. Another type of experiment gives more clear-cut results. Sucrose gradient centrifugation analysis of crude extracts prepared under various conditions, and containing ribosomes as 70s and 100s particles or 30s and 50s particles, shows that in all cases the RNase activity is in the top layer of the centrifuge tube. This means that RNase II is loosely bound to ribosomes. Attachment to ribosomes does not inhibit the enzyme activity at all. NEU and HEPPEL measured the enzyme activity released into the medium during formation of spheroplasts, and reported that no RNase II was liberated under conditions where 75 to 90% of RNase I was found in the medium.

The differences in the modes of existence and activities of RNase I and II may reflect differences in their physiological rôles in the cell.

### c) Polynucleotide Phosphorylase

This enzyme was first found in *E. coli* by LITTAUER and KORNBERG in 1957 (LITTAUER and KORNBERG, 1957) following the discovery in *Azotobacter agilis* (GRUNBERG-MANAGO and OCHOA, 1955) and was thought to operate in RNA synthesis. Nowadays, however, it is considered to have an important rôle in degradation rather than synthesis (GRUNBERG-MANAGO, 1963), because, it is not found in most animal tissues and its reaction favors the phosphorolysis reaction at the intracellular level of phosphates. Moreover, this enzyme has no informational mechanism to determine the specific nucleotide sequence in a polymer. Its mode of existence in cells resembles that of RNase II more than RNase I. Thus, first, though it is found chiefly in the soluble fraction of cell extracts, various amounts of activity are also observed associated with the particulate fraction (SPAHR, 1964); second, its activity is not inhibited by its attachment to ribosomes; third, it is not liberated into the medium during spheroplast formation (NEU and HEPPEL, 1964).

### 2. Physiological Role

Unfortunately, no definite conclusions can yet be drawn on this problem. So, in this review available information will be given about the function of RNA-degradative enzymes in cells and the views of the authors will be given based on the results of *in vitro* experiments on biological activities and on indirect *in vivo* observations.

The probable rôles of RNA-degrading enzymes in growing cells can be summarized as follows:

1. Metabolism of RNAs (m-RNA, s-RNA and ribosomal RNA);
2. Protection against penetration by phage RNA;

3. Supply of nutrients by degrading extracellular RNA;

4. Activation of DNA specific endonuclease I by removing RNA.

### a) Roles in Metabolism of RNA

One possibility is of course that intracellular nucleases work only in dead cells to provide sufficient material for nucleic acid synthesis in viable members of the cell population. Another more attractive possibility, however, is that they participate in some essential reactions in nucleic acid metabolism in growing cells. In fact, various experiments support the latter possibility. Thus the breakdown of messenger RNA after use in protein synthesis and the turnover of transfer and ribosomal RNA under limiting growth conditions may be caused by the nucleases mentioned above.

*(α) m-RNA.* During exponential growth of cells, of all the nucleic acids only m-RNA and the three terminal nucleotides of s-RNA appear to be metabolically unstable (MANDELSTAM, 1960). The turnover of the terminal structure of s-RNA is known to be caused by a specific enzyme differing from RNase.

Hence, at first, studies on the metabolism of m-RNA will be considered. The *in vitro* experiment of ARTMAN and ENGELBERG showed that purified RNase I, as well as the ribosomal fraction, have activity to degrade m-RNA (ARTMAN and ENGELBERG, 1964, 1965). These workers tried to demonstrate the participation of RNase I in m-RNA turnover, based on their *in vitro* investigation, and the following three phenomena which had been observed already by other investigaters. (a) There is 0.1 molecule of this enzyme per 70s particle in cell extracts (SPAHR and HOLLINGWORTH, 1961) and this seems to have some correlation with the fact that on average about ten ribosomes form one polyribosome and a single m-RNA molecule is attached to this cluster; (b) The 30s components to which RNase I is attached also combine with messenger RNA (OKAMOTO and TAKANAMI, 1963; WETTSTEIN et al., 1963); (c) RNase I appears to be inactive to ribosomal RNA under physiological conditions (SPAHR and HOLLINGWORTH, 1961).

However, their *in vivo* experimental results are not consistent with the above hypothesis. They prepared spheroplasts according to the method of NEU and HEPPEL (1964) under conditions in which the latter scarcely contained any RNase I activity. After labeling the RNA with a rapid turnover rate by incubating spheroplasts with $^{32}P$ for a short period, they treated the spheroplasts with actinomycin D to stop the synthesis of messenger RNA. The decay of m-RNA was then traced by the decrease in radioactivity in the trichloroacetic acid-insoluble fraction. The half-life of m-RNA was calculated to be 2 min to 3 min by this procedure. This value is comparable to that obtained for $\beta$-galactosidase m-RNA in *E. coli* (NAKADA and MAGASANIK, 1964) and to the average life time of bulk m-RNA in *Bacillus subtilis* (LEVINTHAL et al., 1962) using intact cells.

Moreover GASTELAND isolated a mutant which lacks RNA I activity and used this to clarify the significance of RNA I activity in bacterial cells (GASTELAND, 1966); he measured the mean life time of m-RNA in the same way in the mutant and obtained a value of about 1 min. Consequently, he also concluded that RNase I has no relation to m-RNA metabolism. There are still many points to be clarified about these experiments, such as the kind of RNA which is rapidly labeled under these conditions. So, the m-RNA responsible for the specific enzyme, which has already been examined in detail in normal cells, should be examined as was the case with $\beta$-galactosidase. However, the conclusion that RNase I does not attack m-RNA in the normal state seems to be acceptable.

Another type of experiment showed that a mutant lacking RNase I activity did not differ in its rate of cell growth from normal cells. This result also suggests that there is no relation between RNase I and m-RNA metabolism in growing cells.

Then, what enzymes participate in the degradation of m-RNA in normal cells? SEKIGUCHI and COHEN (1963) first implicated polynucleotide phosphorylase as the enzyme responsible for the degradation of m-RNA. They observed that the rapidly labeled RNA formed after the infection of *E. coli* with T6-phage, i.e. phage specific m-RNA, was chiefly broken down to nucleoside 5'-diphophates and 5'-monophosphates. They reported that the breakdown was mainly caused by polynucleotide phosphorylase for the following reasons: the reaction required a certain amount of inorganic phosphate, nucleoside 5'-diphosphates accumulated as the major product, 6-aza UDP, a known inhibitor of polynucleotide phosphorylase, had a strong inhibitory effect and externally added polynucleotide phosphorylase accelerated the reaction. They also reported that a kind of phosphodiesterase which was bound to ribosomes took part in the degradation of m-RNA in the system. The phosphodiesterase was later shown to be RNase II by SINGER and TOLBERT (1965). Thus, in phage-infected cells, polynucleotide phosphorylase and RNase II seem to be the enzymes responsible for m-RNA breakdown.

A remarkable characteristic of polynucleotide phosphorylase is that it gives nucleoside 5'-diphosphates as final products. As reduction of ribonucleotides to their corresponding deoxy derivatives in cell-free extracts supplemented with NADPH has been shown to take place at the diphosphate level (REICHARD and RUTBERG, 1960; REICHARD et al., 1961; REICHARD, 1962), this route of degradation appears to be the most effective one when the degradation products are to be reutilized for DNA synthesis. In fact, in phage-infected cells, the [14]C-labeled messenger RNA has been shown to be transferred to phage DNA.

ANDOH et al. (1963) reported a similar observation on the degradation of m-RNA in normal cells under conditions where only m-RNA was selectively degraded. BARONDES and NIRENBERG (1962) showed that polynucleotide phosphorylase was involved in the breakdown of poly U utilized as m-RNA in a cell-free system for protein synthesis.

Recently FUTAI et al. (1966) reported a systematic study of the mode of degradation of m-RNA in a cell-free system where selective depolymerization of m-RNA occurred. They paid special attention to the concentrations of magnesium, potassium, and phosphates, which are known to have great effects on the activity of RNA degrading enzymes. When ribosomes and m-RNA are incubated in Tris buffer containing 5 m$M$ MgCl$_2$ and no other ions, depolymerization of RNA scarcely takes place. In the presence of potassium and phosphates, however, m-RNA becomes susceptible to the action of depolymerizing enzymes. The products are exclusively nucleoside 5'-diphosphates and 5'-monophophates and no nucleoside 3'-phophates are detected. This is also true when the supernatant is employed with the ribosomes as a source of enzymes. This result is consistent with the idea that degradation of m-RNA is due to both RNase II and polynucleotide phosphorylase. The effects of the two enzymes are found to be additive in a cell-free system. Which enzyme actually works *in vivo* may depend upon the source strain and the environmental or cellular conditions.

This suggests that the enzyme activities are controlled by various factors, such as the phase of cell growth, the concentration of ions in the medium and the amounts of the products in the cells.

FUTAI et al. (1966) also used a cell-free system from phage-infected cells and showed that the degradation of m-RNA occurred even in the absence of potassium and phosphates and was especially much increased by addition of the supernatant, in contrast to the result with normal cells. Moreover, considerable amounts of nucleoside 3'-phosphates can be detected as degradation products. This suggests that phage infection induces a disturbance in the control system of the cells and consequently activates RNase I, which may participate in the breakdown of m-RNA. Thus, it seems to us that polynucleotide phosphorylase and RNase II are active in m-RNA turnover in normal cells and RNase I may also participate in the degradation under special conditions. As the turnover of m-RNA is an essential reaction for the growth of cells, it will be strongly influenced by environmental changes. Therefore, the best combination of enzymes for RNA degradation must participate in the breakdown of m-RNA and be influenced by variations in conditions inside and outside the cells.

Enzymes of the type of RNase I are known to be widely distributed in almost all microbes investigated and polynucleotide phosphorylases have also been found in various kind of microorganisms. If RNase II is an important enzyme in the turnover of m-RNA, it must be present in other microorganisms besides *E. coli*. A kind of potassium-activated phosphodiesterase has been found in *Lactobacillus casei* (KEIR et al., 1964). The possible existence of similar enzymes in other microorganisms must be investigated.

*(β) Ribosomal RNA.* Though the turnover of ribosomal RNA has not been fully investigated, this kind of RNA is thought to be completely conserved in normal cells as well as in virus-infected cells (BRENNER et al., 1961). But it is damaged under limited growth conditions, as in the state of so-called maximum concentration (ANDOH and MIZUNO, 1963) of cells, where cells divide slowly because of unfavorable environmental conditions, such as spacial limitation and limitation in the amount of phosphorus (HORIUCHI et al., 1959; MARUYAMA and MIZUNO, 1965) or magnesium (NATORI et al., 1966). When the cells of a temperature-sensitive strain are grown at higher temperature (40 °C), the degradation of ribosomal RNA is also observed (NOZAWA et al., 1967). It is interesting to observe in this case the occurrence of ribosomal particles of abnormally small size which seem to be intermediates of degradation. The particles contain smaller amounts of RNAs and have less ability to adsorb RNase than normal ribosomes. Even under the limiting growth conditions mentioned above, the viability of the cells is not affected and cells can still adapt to form inducible enzymes when inducers are added to the medium, though the ribosomes are obviously damaged (NEIDTHARDT, 1964). The degradation of ribosomal RNA can also be observed when the cells are treated with mitomycin C or colicine E2 (KERSTAN and RAUEN, 1963; NOSE et al., 1966).

The mechanism of ribosomal RNA degradation has been examined extensively in the state of phosphorus deficiency. Bacteria are grown for several generations in phosphate-deficient medium containing a small amount of $^{32}P$ and the changes in the distribution of $^{32}P$ in RNA and its degradation products are analysed. During this stage, the content of ribosomal RNA per cell decreases markedly, though the amounts of other kinds of RNAs (transfer RNA and messenger RNA) are constant.

The products of degradation are a mixture of oligonucleotides. The average size of these oligonucleotides decreases with time after phosphate exhaustion and, at the same time, the base ratio changes from adenine-rich to guanine-rich. The products are a mixture of 3'-phosphate and 2',3'-cyclic phosphate. The proportion of cyclic phosphates decreases with time. These properties are strikingly like those of the products formed by RNase I digestion *in vitro*. This suggests that under limiting growth conditions, ribosomal RNA is degraded by the action of RNase I. Of course, the turnover of messenger RNA or perhaps soluble RNA may occur in this state but, for some reason, the behavior of other RNAs may be masked. Perhaps the amount of ribosomal RNA predominates or the rate of degradation of ribosomal RNA is more rapid than that of messenger RNA in this specific environment. The inorganic phosphates thus obtained are shown to be reutilized for the synthesis of DNA and m-RNA and cells continue to grow under the limiting conditions (MARUYAMA and MIZUNO, 1966).

The investigations so far reported all seem to indicate that RNase I participates in the breakdown of ribosomal RNA when this occurs but does not act on m-RNA in the normal state.

So far, only experiments on wild type cells have been considered. In case of some bacteria, however, even wild type strains are known to lack ribosomal RNase and recently mutant strains of *E. coli* without this type of enzyme activity have been discovered.

The remarkable characteristics of the mutant of *E. coli* which lacks RNase I is the stability of the ribosomes *in vitro* (GASTELAND, 1966). The ribosomes of the mutant cells are not disrupted even when they are put in magnesium-deficient medium. They are not affected by treatment with EDTA or dialysis against urea solution. Moreover, ribosomal RNAs are stable on heat treatment as shown by the unusual reversibility of the melting curve of ribosomes. This seems to indicate that RNase I is concerned with the stability of ribosomes, that is degradation of ribosomal RNA. However, when cells lacking RNase I are subjected to conditions where depolymerization of ribosomal RNA takes place as mentioned above, what enzymes act in place of RNase I?

The first bacteria which were found to lack ribosomal "latent" RNase were *Pseudomonas fluorescens* (WADE and ROBINSON, 1963), and two more strains have been found in the last two years. One is *Pseudomonas aeruginosa* (GRONLUND and CAMPBELL, 1965) where polynucleotide phosphorylase has been shown to be responsible for degradation of ribosomal RNA. The other is *Alkaligenes faecalis* (NATORI et al., 1967). In this strain, ribosomes are very stable in the presence of EDTA, but on addition of potassium and phosphates its RNA-degradation is greatly stimulated and nucleoside 5'-diphosphates and 5'-monophosphates are formed. This means that in this case RNase II and polynucleotide phosphorylase attack the RNA in the ribosomes. However, this was an *in vitro* experiment and caution is necessary in extending the findings to the *in vivo* system.

The fact that degradation of ribosomal RNA even occurs in bacteria which have no detectable RNase I activity does not imply that the enzyme is not important in RNA metabolism. On the contrary, we believe that because the depolymerization of ribosomal RNA is indispensable under special conditions, such as on depletion of the medium of nutrient materials, the reaction is caused by other enzymes even when the route is not advantageous from the point of view of cell economy.

*(γ) s-RNA.* Now the action of RNA-degrading enzymes on soluble RNA will be considered.

It was demonstrated that when ribosomes were treated with soluble RNA and then incubated in the presence of EDTA and phosphates, the bound s-RNA could be hydrolyzed without appreciable destruction of endogenous ribosomal RNA. Not only s-RNA, but also other types of externally added RNA have the same protective effect on ribosomal RNA. In this connection, it is of interest that s-RNA is bound to 50s particles in the protein-synthesizing system and has a double-stranded hairpin structure which is rather resistant to RNA-degrading enzymes other than RNase I, namely, potassium-activated phosphodiesterase and polynucleotide phosphorylase. Polynucleotide phosphorylase, for instance, can digest s-RNA only to the extent of 20 to 50% under conditions where commercial RNA and synthetic polyribonucleotides of single-stranded configuration are almost quantitatively depolymerized (OCHOA, 1957). s-RNA as well as r-RNA is thought to be stable in cells in the normal state. Anyhow, when degradation of s-RNA occurs, the enzyme which attacks s-RNA may well be RNase I, judging from the great effect of secondary structure on the activities of polynucleotide phosphorylase and RNase II[1].

s-RNA may combine with ribosomal RNA through a covalent bond when it enters into the system of protein synthesis on ribosomes (BLOEMANDED et al., 1960). If so, the RNase may function in transnucleotidation between two kinds of RNAs and in the removal of s-RNA from ribosomes.

In considering the function of RNase I in cells it is important to remember that the degradation products of this enzyme are nucleoside 3'-phosphates formed *via* 2',3'-cyclic phosphates (SPAHR and HOLLINGWORTH, 1961; TAL and ELSON, 1963). So far all the enzymes known to participate in the synthesis of nucleic acids *in vivo* require nucleoside 5'-phosphates as substrates. If the breakdown products of RNase I are reutilized for the synthesis of nucleic acids, the digestion products have to be transformed into 5'-nucleotides. One possible advantage of producing 2',3'-cyclic phosphates and 3'-phosphates rather than 5'-phosphates is that cyclic phosphates may be more permeable or more easily transferred into cells than other nucleotides because they are less negatively charged. The observation that the rate of hydrolysis of cyclic phosphates to the corresponding 3'-phosphates by this type of enzyme is much slower than that of the preceding reaction of degradation of RNA (SPAHR and HOLLINGWORTH, 1961) suggests the possibility of accumulation of these components in the cells and their transfer through cellular compartments. Recently, an enzyme named cyclic phosphodiesterase was discovered (ANRAKU, 1964) which catalyzed the hydrolysis of 2',3'-cyclic phosphates more rapidly than ribonucleases So this enzyme can attack cyclic phosphates, if necessary. The 3'-nucleotides thus formed will be converted into nucleosides and inorganic phosphates by the action of phosphatase and these compounds will enter as substrates into the pathway of nucleic acid synthesis. For example, in the phosphate-deficient state, alkaline phosphatase is induced and this attacks 3'-nucleotides so as to supply inorganic phospha-

---

[1] However, the recently found RNase III has to be taken into consideration in this connection.

tes to the reactions required for cell viability (HORIUCHI et al., 1959; TORRIANI, 1960).

Another possibility is based on the fact that RNases such as RNase I may catalyze the synthesis of polymers from nucleoside 2′,3′-cyclic phosphates as shown in RNase T$_1$ (HAYASHI and EGAMI, 1963). This reverse reaction does not require any external energy source. So RNase I may participate in the elongation of the polyribonucleotide chain at the 5′-terminal and in the synthesis of polyribonucleotides with unknown physiological significances, such as in maintainning the equilibrium of polymers and monomers in cells or serving as a polynucleotide pool. In this connection, the existence of RNases with high phosphotransferase activity, but with slight or no hydrolytic activity is noteworthy.

If pool polynucleotides are actually present in cells, they should have no specific nucleotide composition or sequence and hence have no rigid configuration but exist as random coils. Therefore, the polyribonucleotides should be sensitive to RNase II and the degradation products, nucleoside 5′-monophosphates, will be reused for nucleic acid synthesis. Polynucleotide phosphorylase will not act directly on polymers because they have phosphates at the 3′-hydroxyl ends. It could, however, act on them after the elimination 3′-terminal phosphates by phosphomonoesterase. However, there is no evidence for the existence of such a polynucleotide pool or for the elongation of the polynucleotide chain at the 5′-terminal in living cells.

On the basis of the experimental results and speculations mentioned in this chapter, the mode of participation of RNA-degrading enzymes in the metabolism of RNA in cells may be summarized as shown in the following chart.

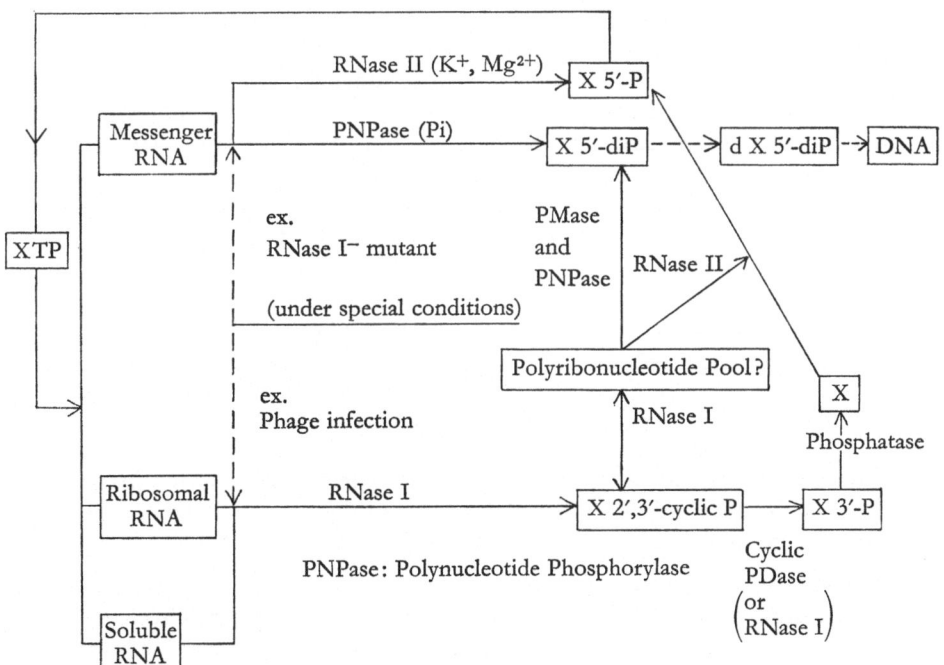

Chart V-1

In conclusion, studies on the physiological rôles of RNA-degrading enzymes in microorganisms are only just beginning and further systematic studies are necessary. It should be added here that, although we have discussed the probable interactions of well-defined RNAs (r-RNA, s-RNA and messenger RNA) and the RNA-degrading enzymes so far known, the possibility that other RNA-degrading enzymes and RNAs will be discovered has to be taken into consideration. For this reason also, the proposed scheme is only tentative.

### b) Other Probable Roles

As summarized at the beginning of this section, RNases probably have other rôles which must be considered. These are (1) protection against penetration of phage RNA, (2) supply of nutrients by degrading extracellular polyribonucleotides and (3) activation of DNA-specific endonuclease I by removing inhibitory RNA.

Unfortunately, there have been few investigations which clarify these possibilities. Here we should remember the existence of RNase I between the cell wall and cell membrane. This suggests that one function of RNase I is to prevent the penetration of external RNA and to depolymerize RNA in the surrounding medium for the supply of digestion products as nutrients. GASTELAND compared the efficiencies of infection of spheroplasts from wild type cells and mutant cells lacking RNase I activity, by RNA from R 17 phage, anticipating results supporting the participation of RNase I in preventing penetration of external RNA. However, no evidence for this was obtained.

The DNA-specific endonuclease I discovered by LEHMAN and ROUSSOS (1962) was shown to be completely inhibited by the addition of a small amount of RNA. The DNase is undetectable in fresh cell extracts, probably because it combines with endogenous RNA. Although anionic polymers other than RNA cannot serve as inhibitors, RNA from different sources such as other bacteria, animal tissues and TMV, can inhibit the enzyme activity. s-RNA and synthetic polyribonucleotides with a double-stranded structure are most effective inhibitors (LEHMAN, 1963). It has been shown in *in vitro* experiments that incubation with high levels of RNase I as well as pancreatic RNase causes an increase in endonuclease activity. In cells, too, RNase I may activate DNase because among intracellular RNA-degrading enzymes RNase I has the highest activity towards RNA with ordered configuration.

The probable participation of RNase III in protection against double-stranded RNA and in activation of DNA specific endonuclease I by removing double-stranded RNA has to be considered.

### 3. Control Mechanism of RNase Activity

How are the activities of the enzymes of RNA degradation controlled in the cell? Several possible mechanisms may be considered on the basis of *in vitro* studies and indirect observations.

1. The existence of specific inhibitors for the enzymes.
2. The protection of RNA by some cellular components or structures.
3. Limitation of enzyme activity owing to the substrate specificity of the enzymes.
4. The intracellular location of the enzymes and substrates.
5. The concentrations of the substrates, products, and effectors.

1. Inhibitors of a number of metabolic enzymes in bacteria have been reported. Recently, proteins which inhibit RNA depolymerases have been found in several microbes including *Neurospora crassa*, (ISHIKAWA et al., personal communication), *Aspergillus oryzae* (UOZUMI and ARIMA, 1966) and *Bacillus subtilis* (KERRETAL, 1965). In *N. crassa*, the RNase fractions obtained from extracts are activated by incubation at 37 °C for 2 to 5 h or by digestion with subtilopeptidase. The activation process has been shown to cause a reduction of molecular weight of the RNase fractions, as estimated by gel filtration, probably by dissociation of "inhibitory" substances. In the case of *Aspergillus oryzae*, an inhibitor of nuclease which is active for DNA as well as RNA is detected in freshly prepared extracts and is observed to disappear or be inactivated during autodigestion. The inhibitor is prepared from the supernatant of extracts which have been heated for 10 min at 80 °C by the methods of precipitation by ammonium sulfate and fractionation on a DEAE-cellulose column. It is nondialyzable, heat-stable and inactivated by the action of protease. The inhibitor combines with the nuclease of *Aspergillus oryzae* stoichiometrically, but does not act on either pancreatic RNase or RNase $T_1$ and $T_2$. Such inhibitors may play a regulatory rôle in cell metabolism, as has been shown to be the case with other enzyme systems (SWARTZ et al., 1956).

RNase I is well known to be latent when it is bound to ribosomes. In this case ribosomes may be regarded as inhibitors of RNase I. In this connection, it is very interesting that the 30s components of ribosomes can adsorb externally added, purified RNase I (SPAHR and HOLLINGWORTH, 1961; ANRAKU and MIZUNO, 1965, 1967). The maximum amount of RNase I adsorbed is about 10 times the amount of RNase originally present on ribosomes in the cells. As the purified ribosomes are shown to contain about 0.1 mole of RNase per mole of 70s particle, it is roughly estimated that ribosomes have the latent ability to adsorb an average of 1 mole of RNase per 70s particle. Commercial bovine pancreatic RNase A and RNase $T_1$ are not adsorbed at all. That is, the mechanism of adsorption of RNase on 30s particles must involve a species-specific reaction.

2. The ribosomal structure may help to protect RNA from enzymatic degradation. Ribosomal RNA is resistant to enzyme attack, being bound to basic protein in the ribosomes. In an incubation mixture containing ribosomes and RNA, it was shown that self-digestion of ribosomes was almost completely suppressed, while most externally added RNA was hydrolyzed (NEU and HEPPEL, 1964). Moreover, m-RNA has been shown to be more susceptibile to degradative enzymes added from outside than ribosomal RNA, probably because it is not bound to protein.

3. Polynucleotide phosphorylase and RNase II are known to be much less active with RNA of ordered configuration, like s-RNA, than with randomly coiled RNA.

4. An important possible mechanism of contrôl is the intracellular location of RNAs and the enzymes which attack them. RNase I is concentrated near the surface of cells in the logarithmic phase and becomes bound to ribosomes in the stationary phase. The shift of RNase with age of the cell seems to have some relationship to the observation that degradation of ribosomal RNA begins when the cells reach a maximum concentration in the medium. It is highly probable that compartmentation participates in regulating RNase activity. SCHLESSINGER (1963) observed that in *Bacillus megaterium* membrane-bound polyribosomes are protected from nuclease attack. SEKIGUCHI and COHEN (SEKIGUCHI and COHEN, 1963) showed that phage-

specific m-RNA was rather resistant to enzyme action just after injection of phage DNA but later became susceptible. This is consistent with the consideration that after use m-RNA is unstable *in vivo*, when it is actively used. In this connection, BEN-HAMIDA and SCHLESSINGER (1966) et al. observed, in a study of RNA breakdown in nitrogen-starved cells, that the onset of degradation of ribosomal RNA was directly linked to a decrease in RNA synthesis and they suggested that this might coincide with the dissociation of polyribosomes. They said that the newly formed m-RNA and ribosomes seemed to be protected from nucleases in polyribosomes in cells in the exponential phase of growth.

5. Finally, the concentration and distribution of substrates, digestion products and effectors within cells must of course be taken into account. In summary, it may be said that the various regulation mechanisms mentioned above will function in accordance with the intracellular and environmental conditions of the cells and will be controlled by the equilibrium of substances in the cells.

We have discussed the physiological aspects of intracellular RNA-degrading enzymes. Special reference has been made to *E. coli*, because so far more or less systematic studies have only been carried out on this organism. Before any general conclusions can be drawn, more extensive studies must be performed on various other kinds of microorganisms.

## C. Role of Extracellular Enzymes

Besides intracellular RNases, extracellular RNases have also been found in culture media of various microorganisms. However, most studies on these have been carried out with an interest in comparative protein chemistry or nucleic acid chemistry and there are few studies on their biological characters.

NISHIMURA and NOMURA (1959) investigated the mode of formation of extracellular RNase in *Bacillus subtilis* strain H. The RNase activity in the medium increases remarkably when growth enters the stationary state and then continues to increase at a constant rate for a certain period. During this period intracellular RNase activity remains less than 0.27% of that in the medium. Essentially the same phenomenon has been observed with *Ustilago zeae* (YANAGIDA et al., 1964), *Ustilago sphaerogena* (ARIMA et al., 1967) and *Neurospora crassa* (TAKAI et al., 1967). The release of extracellular enzymes in the stationary phase of growth has been observed not only with RNases but also with many other extracellular enzymes, and may be regarded as a rather general characteristic of extracellular enzymes of microorganisms.

The extracellular RNase of *Bacillus subtilis* is quite different from the intracellular RNase in its optimum pH, heat stability, ion requirement etc. (NISHIMURA and MARUO, 1960). Moreover, the anti-serum for the intracellular RNase does not react with the extracellular RNase. In contrast, one of the intracellular RNases (RNase $N_3$) of *Neurospora crassa* is so similar to the extracellular RNase (RNase $N_1$) that the former may be regarded as a precursor of the latter (TAKAI et al., 1967).

The yield of extracellular RNases depends largely on the culture conditions. However, it seems likely that they are synthesized by the normal process of protein synthesis through the information transfer: DNA → RNA → protein. This idea is supported by the observation that actinomycin D, which is considered as a specific inhibitor of messenger RNA synthesis, inhibits the synthesis of extracellular RNases.

So far the substances which have been found to have most effect on the yield of certain extracellular RNases are RNA and related compounds. GLITZ and DEKKER (1964) found that RNase accumulates in the culture medium of *Ustilago sphaerogena*, only when RNA is used as the sole carbon source. They consider that extracellular RNase is an inducible enzyme. With this organism the RNase activity in the medium appears to vary closely with the stage of growth. It seems that RNA "induces" RNase formation and the RNase formed digests RNA, yielding directly utilizable carbon compounds, and this results in cell growth. However, as described in Chapter IV, ARIMA et al. found four extracellular RNases (RNases $U_1$, $U_2$, $U_3$ and $U_4$) in the culture medium of the same organism and only two of these (RNase $U_1$ and $U_4$) were "inducible" by RNA. There is some uncertainty about the definition of the term "inducible". However, since increased formation of RNases $U_1$ and $U_4$ also occurs in resting cells and since it is inhibited by chloramphenicol, it is highly probable that RNA "induces" the *de novo* synthesis of the enzymes.

Similar increased formation of a nRNase (guanyloribonuclease) has been observed with *Ustilago zeae* (YANAGIDA et al., 1964). In this case the increased formation of RNase is "induced" not only by RNA, but also by poly U, which cannot be a substrate for the enzyme. By analogy with enzyme induction in the strict sense, such as the induction of $\beta$-galactosidase in *E. coli*, YANAGIDA et al. regarded this increased formation as induction. However, it must be remembered that induction by low-molecular compounds, such as $\beta$-galactosides, and increased formation of enzymes by high-molecular compounds, such as RNA, may be quite different phenomena. The mechanism of the latter requires further investigation and comparison with the former.

What are the physiological rôles of these extracellular RNases? As shown in Chapter II, the RNases so far found in microorganisms may be classified into three categories: non-specific RNases, purine-specific RNases, and guanine-specific RNases. Generally speaking, non-specific RNases have higher molecular weights and are mainly found in microbial cells, while base-specific RNases have lower molecular weights and are mainly found in the culture medium.

If these enzymes have any physiological rôle, it seems most likely that they contribute to the digestion of polyribonucleotides in the medium to give rise to diffusible nutrients. The "induced" formation of RNase by *Ustilago sphaerogena* (GLITZ and DEKKER, 1964) which results in cell growth as mentioned above supports this idea. In this connection it seems noteworthy that microbial cells produce easily diffusible low-molecular RNases which are specific for guanine or purine nucleotides to digest extracellular RNA. As is well known, guanine-rich nucleotides tend to form aggregates (ISHIKURA, 1962; LIPSETT and HEPPEL, 1963). The digestion products of these enzymes without guanylic acid phosphodiester bonds have less tendency to form aggregates and may diffuse through the cell membrane. However, it seems unlikely that extracellular RNases have any such general physiological rôle. Indeed, a phenomenon quite different from that observed with *Ustilago sphaerogena* has been observed with *Neurospora crassa* (TAKAI et al., 1967), namely, addition of RNA instead of phosphate or adenine to the culture media of a wild strain or an adenine-requiring mutant did not "induce" any extracellular RNase. Experimental results suggest that it is not RNase but PDase and PMase that participate in the supply of phosphate or adenosine required for cell growth.

# Chapter VI

# Concluding Remarks

Although microbial RNases are considered valuable materials for studies on protein chemistry, nucleic acid chemistry, comparative biochemistry and cytochemistry, little information about them is available as yet. There are many reports on the existence of "RNases", or more exactly RNA-degrading enzymes, in a wide variety of species of microorganisms. However, unfortunately, most of these only describe the existence and sometimes the partial purification of the enzymes and not their specificities, modes of action, or chemical nature.

Extensive investigations on microbial RNases are very greatly needed.

*Acknowledgement.* The authors are much indebted to Dr. E. OHMURA for his collaboration in the preparation of Chapter III.

# References

ABROSIMOVA-AMELYANCNIK, N. M., R. I. TATARSKAYA, T. V. VENKSTERN, V. D. AKSEL'ROD, and A. A. BAEV: Specificity of Guanyl-RNase of Actinomycetes. Biokhimiya 30, 1269—1276 (1965); Biochemistry (U.S.S.R.) 30, 1086—1092 (1965).

ANDERSON, T. H., and C. E. CARTER: Acid-soluble ribosomal ribonuclease of *Escherichia coli*. Biochemistry 4, 1102—1107 (1965).

ANDOH, T., and D. MIZUNO: Turnover of nucleic acid and of protein in the maximum concentration of *E. coli* with special reference to messenger RNA on ribosomes. J. Biochem. (Tokyo) 54, 237—245 (1963).

—, S. NATORI, and D. MIZUNO: The degradation of *Escherichia coli* messenger RNA by polynucleotide phosphorylase. J. Biochem. (Tokyo) 54, 339—348 (1963); Biochim. et Biophys. Acta 76, 477—479 (1963).

ANRAKU, Y.: A new cyclic phosphodiesterase having a 3'-nucleotidase activity from *Escherichia coli* B. I. purification and some properties of the enzyme. J. Biol. Chem. 239, 3412—3419 (1964); II. Further studies on substrate specificity and mode of action of the enzyme. J. Biol. Chem. 239, 3420—3424 (1964).

—, and D. MIZUNO: A ribonuclease from the debris of *Escherichia coli*. Biochem. Biophys. Res. Commun. 18, 462—468 (1965).

— — Comparative study on the ribonuclease isolated from the debris and ribosome fraction of *Escherichia coli* J. Biochem. (Tokyo) 61, 70—80 (1967).

— — Ribonuclease-cyclic phosphodiesterase system in *Escherichia coli*. J. Biochem. (Tokyo) 61, 81—88 (1967).

ARIMA, T., T. UCHIDA, and F. EGAMI: Studies of extracellular ribonucleases of *Ustilago Sphaerogena*. I. Purification and properties. Biochem. J. 106, 601—608 (1968); II. Characterization of substrate specificity with special reference to purine-specific ribonuclease. Biochem. J. 106, 609—613 (1968).

ARTMAN, M., and H. ENGELBERG: Degradation of rapidly-turned-over ribonucleic acid to acid-soluble compounds by *Escherichia coli* ribosomes. Biochim. et Biophys. Acta 80, 517—520 (1964).

— — Breakdown of rapidly labelled ribonucleic acid in actinomycin-sensitive spheroplasts of *Escherichia coli* devoid of ribonuclease. Biochim. et Biophys. Acta 95, 687—689 (1965).

AZEGAMI, M., and K. IWAI: Specific modification of nucleic acids and their constituents with trinitrophenyl group. J. Biochem. (Tokyo) 55, 346—348 (1964).

BOLTON, E. T., R. J. BRITTEN, D. B. COWIE, B. J. ACCARTHY, K. MCQUILLEN, and R. B. ROBERTS: Carnegie Inst. of Washington, Ann. Rep. (1959).

BARONDES, S. H., and M. W. NIRENBERG: Fate of synthetic polynucleotide directing cell-free protein synthesis. I. characteristics of degradation. Science 138, 810—813 (1962).

BEN-HAMIDA, F., and D. SCHLESSINGER: Synthesis and breakdown of ribonucleic acid in *Escherichia coli* starving for nitrogen. Biochim. et Biophys. Acta 119, 183—191 (1966).

BLOEMANDEL, H., L. BOSCH, and M. SLUYSER: Studies on cytoplasmic ribonucleic acid from rat liver II fractionation and function of microsomal ribonucleic acid. Biochim. et Biophys. Acta 41, 454—461 (1960).

BRENNER, S., F. JACOB, and M. MESELSON: An unstable intermediate carrying information from genes to ribosomes for protein synthesis. Nature 190, 578—581 (1961).

COLEMAN, G., and W. H. ELIOTT: Extracellular ribonuclease formation in *Bacillus subtilis* and its stimulation by actinomycin D. Biochem. J. 95, 699—706 (1965).

CONTI, S. F., and M. E. GETTNER: Electron microscopy of cellular division in *Escherichia coli*. J. Bacteriol. 83, 544—550 (1962).

CUNNINGHAM, L., B. W. CATLIN, and M. PRIVAT DE GARILHE: A deoxyribonuclease of *Micrococcus pyogenes*. J. Am. Chem. Soc. 78, 4642—4645 (1956).

EGAMI, F.: Studies on ribonuclease $T_1$ (in Japanese) J. Chem. Soc. Japan 87, 909—917 (1966).
— Chemical nature and mode of action of ribonuclease $T_1$. J. Sci. Ind. Research (India) 25, 442—449 (1966).
—, K. TAKAHASHI, and T. UCHIDA: Ribonucleases in Takadiastase: chemical nature and applications, In: Prog. in nucleic acid res. and mol. biol. edited by DAVIDSON, J. N., and W. E. COHN. New York and London: Academic Press 3, 59—101 (1964).
ELSON, D.: Latent ribonuclease activity in a ribonucleaprotein. Biochim. et Biophys. Acta 27, 216—217 (1958).
— Latent enzymic activity of a ribonucleoprotein isolated from *Escherichia coli*. Biochim. et Biophys. Acta 36, 372—386 (1959).
—, and M. TAL: Biochemical differences in ribonucleoproteins. Biochim. et Biophys. Acta 36, 281—284 (1959).
FUTAI, M., Y. ANRAKU, and D. MIZUNO: The roles of three enzymes in messenger RNA degradation in cell-free systems from normal or phage-infected *Escherichia coli*. Biochim. et Biophys. Acta 119, 373—384 (1966).
GASTELAND, R. F.: Isolation and characterization of ribonuclease I mutants of *Escherichia coli*. J. Mol. Biol. 16, 67—84 (1966).
GLITZ, D. G., and C. A. DEKKER: Studies on a ribonuclease from *Ustilago sphaerogena*. I. Purification and properties at the enzyme; Biochemistry 3, 1391—1399, (1964). II. Specificity of the enzyme Biochemistry 3, 1399—1406 (1964).
GRONLUND, A. F., and J. J. R. CAMPBELL: Enzymatic degradation of ribosomes during endogenous respiration of *Pseudomonas aeruginosa*. J. Bacteriol. 90, 1—7 (1965).
GRUNBERG-MANAGO, M.: Polynucleotide phosphorylase. In: Progress in nucleic acid research (DAVIDSON, J. N., and W. E. COHEN, eds.) 1, 93—133 (1963).
—, and S. OCHOA: Enzymatic synthesis and breakdown of polynucleotides; polynucleotide phosphorylase. J. Am. Chem. Soc. 77, 3165—3166 (1955).
HARTLEY, R. W. JR., G. W. RUSHITZKY, A. E. GRECO, and H. A. SOBER: Studies on B. *subtilis* ribonuclease. II. Molecular weight and physical homogeneity. Biochemistry 2, 794—797 (1963).
HAYASHI, H., and F. EGAMI: Fractionation and properties of guanylic acid polymers synthesized by ribonuclease $T_1$. J. Biochem. (Tokyo) 53, 176—180 (1963).
HENDLER, R. W., W. G. BANFIELD, J. TANI, and E. L. KLUFF: On the cytological unit for protein synthesis *in vivo* in *E. coli*. III. Electron microscopic and ultracentrifugal examination of intact cells and fractions. Biochim. et Biophys. Acta 80, 309—314 (1964).
HIRAMARU, M., J. SOKAWA, T. UCHIDA, and F. EGAMI: Action of RNase $T_1$ on RNA methylated with dimethyl sulfate (in Japanese), presented at the Japan Biochem. Congress (1966). Seikagaku 38, 662 (1966).
HOLLEY, R. W., J. APGAR, G. A. EVERETT, J. T. MADISON, M. MARQUISEE, S. H. MERRILL, J. R. PENSWICK, and A. ZAMIR: Structure of a ribonucleic acid. Science 147, 1462—1465. (1965).
HORIUCHI, T., S. HORIUCHI, and D. MIZUNO: Degradation of ribonucleic acid in *Escherichia coli* in phosphorus-deficient culture. Biochim. et Biophys. Acta 31, 570—572 (1959).
— — — A possible negative feedback phenomenon controlling formation of alkaline phosphomonoesterase in *Escherichia coli*. Nature 183, 1529—1530 (1959).
IRIE, M.: Enzymatic depolymerization of synthetic polynucleotides, poly A, poly C and poly U by ribonuclease $T_1$ preparations. J. Biochem. (Tokyo) 58, 599—603 (1965).
— Inhibition of ribonuclease $T_1$ by various kinds of nucleotides. J. Biochem. (Tokyo) 56, 496—497 (1964).
— Isolation and properties of a ribonuclease from *Aspergillus saitoi*. J. Biochem. (Tokyo) 62, 509—518 (1967).
ISHIKURA, H.: Further studies on the fractionation of ribonuclease I-core by gel filtration. J. Biochem. Tokyo 52, 324—332 (1962).
ISHIKAWA, T., K. HASUNUMA, and A. TOH-E: Personal communication.
ITAGAKI, K., Y. KURIYAMA, Y. SHIOBARA, H. HAYASHI, T. YAMAGATA, and F. EGAMI: Oligoribonucleotides effective for streptolysin S' formation (in Japanese). Seikagaku 37, 217—225 (1965).

IWANOFF, L.: Über die fermentative Zersetzung der Thymonucleinsäure durch Schimmel-
pilze. Z. physiol. Chem. 39, 31—43 (1903).

KADOWAKI, K., J. HONDA, and B. MARUO: Effects of actinomycin D and 5-fluorouracil on
the formation of enzymes in *Bacillus subtilis*. Biochim. et Biophys. Acta 103, 311—318
(1965).

KASAI, H., K. TAKAHASHI, and T. ANDO: Structure and function of ribonuclease T₁.
Symposion Enzyme Chem. (Japan) 17, 77—80 (1965).

KASAI, K.: Studies on the reoxidation of reduced rebonuclease T₁. J. Biochem. (Tokyo) 57,
372—379 (1965).

KEIR, H. M., R. H. MATHOG, and C. E. CARTER: Purification of a potassium ion-activated
ribonuclease, 5'-phosphodiesterase from *Lactobacillus casei*. Biochemistry 3, 1188—1193
(1964).

KERR, I. M., E. A. PRATT, and I. R. LEHMAN: Exonucleolytic degradation of high-molecular-
weight DNA and RNA to nucleoside 3'-phosphates by a nuclease from *B. subtilis*.
Biochem. Biophys. Res. Commun. 20, 154—162 (1965)

KERSTEN, H., and H. M. RAUEN: Degradation of deoxyribonucleic acid in *Escherichia coli*
cells treated with mitomycin C. Nature 190, 1195—1196 (1963).

KOCHETKOV, N. K., E. I. BUDOWSKY, N. E. BOUDE, and L. M. KLEBANORA: Specific cleavage
of tRNA at inosine and 2-dimethyl-amino-6-oxopurine residues. Biochim. et Biophys.
Acta 134, 492—495 (1967).

KUNINAKA, A.: Studies on the decomposition of nucleic acid by microorganisms, I. J. Agr.
Chem. Soc. Japan 28, 282—287 (1954).

— Studies on the decomposition of nucleic acid by microorganisms, V. J. Agr. Chem. Soc.
Japan 30, 583 (1956).

— Enzymic degradation of yeast ribonucleic acid and its related compounds by *Aspergillus
oryzae*. J. Gen. Appl. Microbiol. 3, 55—92 (1957).

KURIYAMA, Y., J. KOYAMA, and F. EGAMI: Digestion of chemically synthesized poly G by
RNase T₁ and spleen phosphodiesterase (in Japanese). Seikagaku 36, 135—139 (1964).

LEHMAN, I. R.: The nucleases of *Escherichia coli*. In: Progress in nucleic acid research
(DAVIDSON, J. N., and W. E. COHN, eds.) 2, 83—123 (1963).

—, G. G. ROUSSOS, and E. A. PRATT: The Deoxyribonucleases of *Escherichia coli*. III. Studies
on the nature of the inhibition of endonuclease by ribonucleic acid. J. Biol. Chem. 237,
829—833 (1962).

LEVINTHAL, C., A. KEYNAN, and A. HIGA: Messenger RNA turnover and protein synthesis
in *Bacillus subtilis* inhibited by actinomycin D. Proc. Nat. Acad. Sci. US 48, 1631—1638
(1962).

LIPSETT, M. N., and L. A. HEPPEL: The separation of guanosine oligonucleotides: Use of
urea to avoid aggregate formation. J. Am. Chem. Soc. 85, 118 (1963).

LITTAUER, U. Z., and A. KORNBERG: Reversible synthesis of polyribonucleotides with an
enzyme from *Escherichia coli*. J. Biol. Chem. 226, 1077—1092 (1957).

MADISON, J. T., and R. W. HOLLEY: The presence of 5,6-dihydrouridylic acid in yeast
"soluble" ribonucleic acid. Biochem. Biophys. Res. Commun. 18, 153—157 (1965).

MANDELSTAM, J.: The intracellular turnover of protein and nucleic acids and its role in
biochemical differentiation. Bacteriol. Rev. 24, 289 (1960).

MARUYAMA, H., and D. MIZUNO: The participation of ribonuclease in the degradation of
*Escherichia coli* ribosomal ribonucleic acid as revealed by oligonucleotides accumulation
in the phosphorus-deficient stage. Biochim. et Biophys. Acta 108, 593—604 (1965).

— — Reutilization of degradation products of ribosomal ribonucleic acid in *Escherichia coli*
strain B during the Phosphorus-deficient stage. Biochim. et Biophys. Acta 123, 510—522
(1966).

MASUI, M., Y. HONDA, K. HIKITA, T. IWATA, I-HUNG PAN, and I. TAKI: Studies on nucleo-
polymerases. I. Occurrence of ribonuclease and deoxy-ribonuclease in culture filtrates of
various bacteria. Osaka City Med. J. 7, 141—151 (1956).

MCCARTY, M.: The occurrence of nucleases in culture filtrates of group A hemolytic streptococci. J. Exp. Med. 88, 181—188 (1948).

MCCULLY, K. S., and G. L. CANTONI: On the specificity of takadiastase T₁ ribonuclease.
Biochim. et Biophys. Acta 51, 190—192 (1961).

84                                        References

Minato, S., T. Tagawa, and K. Nakanishi: Purification of ribonuclease $T_1$. Ann. Sankyo
    Res. Lab. **15**, 122—128 (1963).
— — — Crystallization of ribonuclease $T_1$. J. Biochem. (Tokyo) **59**, 443—448 (1966).
Nakada, D., and B. Magasanik: The roles of inducer and catabolite repressor in the syn-
    thesis of $\beta$-galactosidase by *Escherichia coli*. J. Mol. Biol. **8**, 105—127 (1964).
Naoi-Tada, M., K. Sato-Asano, and F. Egami: Studies on ribonucleases in takadiastase.
    III. Purification and properties of ribonuclease $T_2$. J. Biochem. (Tokyo) **46**, 757—764
    (1959).
Natori, S., R. Nozawa, and D. Mizuno: The turnover of ribosomal RNA of *Escherichia
    coli* in a magnesium-deficient stage. Biochim. et biophys. Acta **114**, 245—253 (1966).
—, T. Horiuchi, and D. Mizuno: Absence of ribonuclease in *Alcaligenes faecalis* and a
    possible mechanism of RNA degradation in this bacteria. Biochim. et Biophys. Acta
    **134**, 337 (1967).
Neidthardt, F. C.: The regulation of RNA synthesis in bacteria. In: Progress in nucleic
    acid research (Davidson, J. N., and W. E. Cohn, eds.) **3**, 145—181 (1964).
Neu, H. C., and L. A. Heppel: The release of ribonuclease into the medium when *E. coli*
    cells are converted to spheroplasts. Biochem. Biophys. Res. Commun. **14**, 109—112
    (1964).
— — The release of ribonuclease into the medium when *Escherichia coli* cells are converted
    to spheroplasts. J. Biol. Chem. **239**, 3893—3900 (1964).
— — The release of enzymes from *Escherichia coli* by osmotic shock and during the forma-
    tion of spheroplasts. J. Biol. Chem. **240**, 3685—3692 (1965).
Nikai, M., Z. Minami, T. Yamazaki, and A. Tsugita: Studies on the nucleases of a strain
    of *Bacillus subtilis*. J. Biochem. (Tokyo) **57**, 96—99 (1965).
Nishimura, S.: Extracellular ribonucleases from *Bacillus subtilis*. I. Crystallization and
    specificity. Biochim. et Biophys. Acta **45**, 15—27 (1960).
— *Bacillus subtilis* ribonuclease. In: Procedures in nucleic acid research, p. 56—63, edited by
    Cantoni, G. L., and D. R. Davies. New York and London: Harper and Row, Publ.
    1966.
—, and M. Nomura: Ribonuclease of *Bacillus subtilis*. Biochim. et Biophys. Acta **30**, 430—
    432 (1958).
— — Ribonuclease of *Bacillus subtilis*. J. Biochem. (Tokyo) **46**, 161—167 (1959); Biochim.
    et Biophys. Acta **30**, 430—431 (1958).
—, and B. Maruo: Intracellular ribonuclease from *Bacillus subtilis*. Biochim. et Biophys.
    Acta **40**, 355—357 (1960).
—, and H. Ozawa: Extracellular ribonuclease from *Bacillus subtilis*. II. Its amino acid com-
    position and structural comparison with bovine pancreatic ribonuclease. Biochim. et
    Biophys. Acta **55**, 421—430 (1962).
Noguchi, J.: Über den Abbau von Nucleinsäuren durch Takadiastase. Biochem. Z. **147**,
    255—257 (1924).
Nose, K., D. Mizuno, and H. Ozeki: Degradation of ribosomal RNA from *Escherichia coli*
    induced by colicine $E_2$. Biochim. et. Biophys. Acta **119**, 636—638 (1966).
Nozawa, R., T. Horiuchi, and D. Mizuno: Degradation of ribosomal RNA in a tempera-
    ture-sensitive *Escherichia coli*. Arch. Biochem. Biophys. **118**, 402—409 (1967).
Ochoa, S.: Enzymic synthesis of polynucleotides. III. Phosphorolysis of natural and syn-
    thetic ribopolynucleotides. Arch. Biochem. Biophys. **69**, 119—129 (1957).
Ohtaka, Y., K. Uchida, and T. Sakai: Purification and properties of ribonuclease from
    yeast. J. Biochem. (Tokyo) **54**, 322—327 (1963).
Okamoto, T., and M. Takanami: Interaction of ribosomes and some synthetic poly-
    ribonucleotides. Biochim. et Biophys. Acta **68**, 325—327 (1963).
Otani, H.: Über die nukleinsäurespaltenden Fermente in den Schimmelpilzen. Acta Schol.
    Med. Univ. Kioto **17**, 323—329 (1935).
Pardee, A. B., and R. E. Kunkee: Enzyme activity and bacteriophage infection. II. Ac-
    tivities before and after virus infection. J. Biol. Chem. **199**, 9—24 (1952).
—, and I. Williams: The increase in desoxyribonuclease of virus-infected *E. coli*. Arch.
    Biochem. Biophys. **40**, 222—224 (1952).

PIRIE, N. W.: The isolation from normal tobacco leaves of nucleoprotein with some similarity to plant viruses. Biochem. J. 42, 614—625 (1950).

— Macromolecular nucleoproteins of healthy tobacco leaves. C.A. 52, 1195c (1958), Biochemistry (U.S.S.R.) 22, 133—140 (1957).

POLLOCK, M. R., and M. H. RICHMOND: Low cyst(e)ine content of bacterial extracellular proteins: Its possible physiological significance. Nature 194, 446—449 (1962).

REICHARD, P.: Enzymatic synthesis of deoxyribonucleotides. I. Formation of deoxycytidine diphosphate from cytidine diphosphate with enzyme from *Escherichia coli*. J. Biol. Chem. 237, 3513—3519 (1962).

—, A. BALDESTEN, and L. RUTBERG: Formation of deoxycytidine phosphates from cytidine phosphates in extracts from *Escherichia coli*. J. Biol. Chem. 236, 1150—1157 (1961).

—, and L. RUTBERG: Formation of deoxycytidine 5'-phosphate from cytidine 5'-phosphate with enzymes from *Escherichia coli*. Biochim. et Biophys. Acta 37, 554—555 (1960).

ROBERTSON, H. D., R. F. WEBSTER, and N. D. ZINDER: A nuclease specific for double-stranded RNA. Virology 32, 718—719 (1967).

RUSHIZKY, G. W., and H. A. SOBER: Characterization of the major compounds in ribonuclease T₁ digests of ribonucleic acid. V. mono-, di, and trinucleotides. J. Biol. Chem. 237, 834—840 (1962).

—, A. E. GRECO, R. W. HARTLEY, JR., and H. A. SOBER: Studies on *B. subtilis* ribonuclease. I. Characterization of enzymatic specificity. Biochemistry 2, 787—797 (1963).

— — — — Studies on the characterization of ribonucleases. J. Biol. Chem. 239, 2165—2169 (1964).

—, and H. A. SOBER: Studies on the specificity of ribonuclease T₂. J. Biol. Chem. 238, 371—376 (1963).

SAIGUSA, S., K. TAKAHASHI, and F. EGAMI: Chemical and enzymatic modification and catalytic activity of RNase T₁ (in Japanese). Symposia on Enzyme Chem. (Japan) 15, 48—52 (1961).

SARUNO, R.: Crystallization of ribonuclease from *Aspergillus oryzae*. Bull. Agri. Chem. Soc. Japan 20, 57—58 (1956).

SATO, K., and F. EGAMI: Studies on ribonucleases in Takadiastase. I. J. Biochem. (Tokyo) 44, 753—762 (1957).

SATO-ASANO, K.: Studies on ribonucleases in takadiastase. II. Specificity of ribonuclease T₁. J. Biochem. (Tokyo) 46, 31—37 (1959).

— Studies on ribonucleases in takadiastase. V. Synthetic reaction by ribonuclease T₁. J. Biochem. (Tokyo) 48, 284—291 (1960).

—, and F. EGAMI: Réaction synthétique par la ribonucléase T₁. Biochim. et Biophys. Acta 29, 655—656 (1958).

— — Ribonucleases in takadiastase. Nature 185, 462—463 (1960).

—, and Y. FUJII: Studies on ribonucleases in takadiastase. IV. Action of ribonuclease T₁ on deamino-ribonucleic acid. J. Biochem. (Tokyo) 47, 608—615 (1960).

SATO, S., and F. EGAMI: On the interaction of ribonuclease T₁ and guanosine 2'-phosphate and related compounds. Biochem. Z. 342, 437—448 (1965).

—, T. UCHIDA, and F. EGAMI: Action of ribonuclease T₂ on 2',3'-cyclic nucleotides and related compounds. Arch. Biochem. Biophys. 115, 48—52 (1966).

SCHLESSINGER, D.: Protein synthesis by polyribosomes on protoplast membranes of *Bacillus megaterium*. J. Mol. Biol. 7, 569—582 (1963).

SEKIGUCHI, M., and S. S. COHEN: The selective degradation of phage-induced ribonucleic acid by polynucleotide phosphorylase. J. Biol. Chem. 238, 349—356 (1963).

SHORTMAN, K.: Studies on cellular inhibitors of ribonuclease. II. Some properties of the inhibitor from rat liver. Biochim. et Biophys. Acta 55, 88—96 (1962).

SHIIO, T., K. ISHII, and S. SHIMIZU: Ribonuclease found in insoluble particle fraction from *Azotobacter agilis* (VINELANDII). J. Biochem. (Tokyo) 59, 363—373 (1960).

SHIOBARA, Y., K. TAKAHASHI, and F. EGAMI: The structure and function of ribonuclease T₁. J. Biochem. (Tokyo) 52, 267—271 (1962).

SIEKEVITZ, P., and G. E. PALADE: A cytochemical study on the pancreas of the guinea pig. II. Functional variations in the enzymatic activity of microsomes. J. Biophys. Biochem. Cytol. 4, 309—318 (1958).

SINGER, M. F., and G. TOLBERT: Purification and properties of a potassium-activated phosphodiesterase (RNase II) from *Escherichia coli*. Biochemistry **4**, 1319—1330 (1965).

SMEATON, J. R., W. H. ELLIOTT, and G. COLEMAN: An inhibitor in *Bacillus subtilis* of its extracellular ribonuclease. Biochem. Biophys. Res. Commun. **18**, 36—42 (1965).

SPAHR, P. F.: Purification and properties of ribonuclease II from *Escherichia coli*. J. Biol. Chem. **239**, 3716—3726 (1964).

— The isolation, assay, and properties of ribonuclease I from *Escherichia coli*. In: Procedures in nucleic acid research, p. 64—70, edited by CANTONI, G. L., and D. R. DAVIES. New York and London: Harper and Row, Publ. 1966.

—, and B. R. HOLLINGWORTH: Purification and mechanism of action of ribonuclease from *Escherichia coli* ribosomes. J. Biol. Chem. **236**, 823—831 (1961).

—, and D. SCHLESSINGER: Breakdown of messenger ribonucleic acid by a potassium-activated phosphodiesterase from *Escherichia coli*. J. Biol. Chem. **238**, 2251—2253 (1963).

STAEHELIN, M.: On the specificity of RNase $T_1$. Biochim. et Biophys. Acta **87**, 493—495 (1964).

STENESH, J. J.: Doctoral Thesis, University of Calfornia, Berkeley.

SUHARA, I., E. KUSABA, and E. OHMURA: Acrocylindrium sp. MM 21. Symposia on Enzyme Chem. (Japan) **16**, 115—118 (1964).

SUIT, J. C.: Localization of DNA-like ribonucleic acid in a "membrane" fraction of *Escherichia coli*. Biochim. et Biophys. Acta **72**, 488—490 (1963).

SUSKIND, S. R., and D. M. BONNER: The effect of mutation on ribonucleic acid protein and ribonuclease formation in *Neurospora crassa*. Biochim. et Biophys. Acta **43**, 173—182 (1960).

SWORTZ, M. N., N. O. KAPLAN, and M. E. FRECH: Significance of "heat-activated" enzymes. Science **123**, 50—53 (1956).

TADA, M.: Specifiicity of RNase $T_2$ for minor base nucleotides (abstract). J. Japan. Biochem. Soc. (Seikagaku) **38**, 662 (1966).

TAKAHASHI, K.: The structure and function of ribonuclease $T_1$. I. Chromatographic purification and properties of ribonuclease $T_1$. I. Chromatographic purification and properties of ribonuclease $T_1$. J. Biochem. (Tokyo) **49**, 1—8 (1961); II. Further purification and amino acid composition of ribonuclease $T_1$. J. Biochem. (Tokyo) **51**, 95—108 (1962); III. Amino- and carboxyl-terminal sequences of ribonuclease $T_1$. J. Biochem. (Tokyo) **52**, 72—81 (1962).

— The amino acid sequence of ribonuclease $T_1$. J. Biol. Chem. **240**, PC4117—PC4119 (1965).

— The structure and function of ribonuclease $T_1$. VII. Further investigation on amino acid composition and some other properties of ribonuclease $T_1$. J. Biochem. (Tokyo) **60** 239—245 (1966).

—, S. MOORE, and W. H. STEIN: The reaction of iodoacetate with glutamic acid residue-58 at the active center of ribonuclease $T_1$. 7th Inter. Cong. Biochem. Tokyo, Abstracts IV, 585 (1967).

TAKAI, N., T. UCHIDA, and F. EGAMI: Purification and properties of ribonuclease $N_1$. An extracellular ribonuclease of *Neurospora crassa*. Biochim. et Biophys. Acta **128**, 218—220 (1966).

— — — Ribonuclease, phosphodiesterases and phosphomonoesterase of *Neurospora crassa* in various culture conditions. J. Japan. Biochem. Soc. (Seikagaku) **39**, 285—290 (1967).

TAKEMURA, S., and M. MIYAZAKI: Behavior of ribonucleases $T_1$ and $T_2$ towards riboapyrimidic acids. J. Biochem. (Tokyo) **46**, 1281—1283 (1959).

TAL, M., and D. ELSON: The location of ribonuclease in *Escherichia coli*. Biochim. et Biophys. Acta **76**, 40—47 (1963).

TANAKA, K.: Ribonuclease from *Streptomyces erythreus*. In: Procedures in nucleic acid research, p. 14—19. Edited by CANTONI, G. L., and D. R. DAVIES. New York and London: Harper and Row, Publ. 1966.

—, and G. L. CANTONI: On the specificity of RNase from *Streptomyces erythreus*. Biochim. et Biophys. Acta **72**, 641—642 (1963).

TASHIRO, Y.: Studies on the ribonucleoprotein particles. IV. Spontaneous degradation and ribonuclease activity of the microsomal ribonucleoprotein particles. J. Biochem. (Tokyo) **45**, 937—945 (1958).

TATARSKAYA, R. I., N. H. ABROSIMOVA-AMELYANCNIK, V. D. AKSEL'ROD, A. I. KORENYAKO, N. YA. NIEDRA, and A. A. BAEV: Isolation and purification of guanyl-ribonuclease from *Actinomycetes*. Biokhyimiya **31**, 1017—1025 (1966); Biochemistry (U.S.S.R.) **31**, 882—890 (1966).

TILLET, W. S., S. SHERRY, and L. R. CHRISTENSEN: Stroptococcal desoxyribonuclease: Significance in lysis of purulent exudates and production by strains of hemolytic streptococci Proc. Soc. Exp. Biol. Med. **68**, 184—188 (1948).

TISSIERES, A., and J. D. WATSON: Breakdown of messenger ribonucleic acid during *in vitro* amino acid incorporation into proteins. Proc. Nat. Acad. Sci. US **48**, 1061—1069 (1962).

TORRIANI, A. M.: Influence of inorganic phosphate in the formation of phosphates by *Escherichia coli*. Biochim. et Biophys. Acta **38**, 460—469 (1960).

TS'O, P. O. P., J. BONNER, and J. VINOGRAD: Structure and properties of microsomal nucleoprotein particles from pea seedlings. Biochim. et Biophys. Acta **30**, 570—582 (1958).

UCHIDA, T.: A simplified method for the purification of ribonuclease $T_1$. J. Biochem. (Tokyo) **57**, 547—553 (1965).

— Purification and properties of RNase $T_2$. J. Biochem. (Tokyo) **60**, 115—132 (1966).

—, and F. EGAMI: Ribonuclease $T_1$ from Taka-diastase. In: Procedures in nucleic acid research, p. 3—13. Edited by CANTONI, G. L., and D. R. DAVIES. New York and London: Harper and Row, Publ. 1966.

— — Ribonuclease $T_2$ from Taka-diastase. In: Procedures in nucleic acid research, p. 46—55. Edited by CANTONI, G. L., and D. R. DAVIES. New York and London: Harper and Row, Publ. 1966.

— — The specificity of ribonuclease $T_2$. J. Biochem. (Tokyo) **61**, 44—53 (1967).

UOZUMI, T., and K. ARIMA: Nuclease inhibitor of *Aspergillus oryzae*. J. Japan. Biochem. Soc. (Seikagaku) **38**, 508—509 (1966).

—, G. TAMURA, and K. ARIMA: A new nuclease of *Aspergillus oryzae* and its inhibitor in the cells. 7th Inter. Cong. Biochem. Tokyo, Absts. IV, 758 (1967).

WADE, H. E., and S. LOVETT: Polynucleotide phosphorylase in ribosomes from *Escherichia coli*. Biochem. J. **81**, 319—328 (1961).

—, and H. K. ROBINSON: Absence of ribonuclease from the ribosomes of *Pseudomonas fluorescens*. Nature **200**, 661—663 (1963).

WETTSTEIN, F. O., T. STAEHELIN, and H. NOLL: Ribosomal aggregate engaged in protein synthesis: Characterization of the ergosome. Nature **197**, 430—435 (1963).

WHITFELD, P. R., and H. WITZEL: On the mechanism of action of Takadiastase ribonuclease $T_1$. Biochim. et Biophys. Acta **72**, 338—341 (1963).

WHITFELD, G. W., and H. WITZEL: Specificity of *Bacillus subtilis* ribonuclease. Biochim. et Biophys. Acta **72**, 362—371 (1963).

YAMAGATA, S., K. TAKAHASHI, and F. EGAMI: The structure and function of ribonuclease $T_1$. IV. Photooxidation of ribonuclease $T_1$. J. Biochem. (Tokyo) **52**, 261—266 (1962).

— — — The structure and function of ribonuclease $T_1$. VI. Reduction of disulfide bonds of ribonuclease $T_1$. J. Biochem. (Tokyo) **52**, 272—274 (1962).

YAMAMOTO, M., and J. TANAKA: Optical rotation of RNase $T_1$. Presented at the 20th Cong. of Chem. Soc. Japan, Proceeding of the Congress, p. 688 (1967).

YANAGIDA, M., T. UCHIDA, and F. EGAMI: Culture of *Ustilago zeae* with RNA or poly U as phosphorus source. J. Agr. Chem. Soc. Japan (Nippon Nôgei-kagaku Kaishi) **38**, 531—535 (1964).

YONEDA, M.: Studies on the ribonuclease from *Streptomyces albogriselus*. J. Biochem. (Tokyo) **55**, 469—474 (1964).

# Subject Index

Type-setting and printing: Carl Ritter & Co., Wiesbaden

# Molecular Biology, Biochemistry and Biophysics
# Molekularbiologie, Biochemie und Biophysik